HOME INSPECTOR CONFIDENTIAL

INSIDER SECRETS TO BECOMING A SUCCESSFUL HOME INSPECTOR

INSIDER SECRETS TO BECOMING A SUCCESSFUL HOME INSPECTOR

HOME INSPECTOR CONFIDENTIAL

BY
MATT FELLMAN

INKWATER
PRESS

PORTLAND • OREGON
INKWATERPRESS.COM

Edited by Andrew Durkin
Cover and interior design by Masha Shubin
Vintage Investigator in Front of House © stockasso. Man Searching With
Magnifying Glass © iconisa. BigStockPhoto.com

Publisher: Inkwater Press | www.inkwaterpress.com

ISBN-13 978-1-62901-684-9 | ISBN-10 1-62901-684-5

3 5 7 9 10 8 6 4 2

TABLE OF CONTENTS

PART THREE
PERFORMING AN INSPECTION

PART FOUR

STAYING EDUCATED AND KEEPING
AN EYE ON YOUR COMPETITION

PART FIVE

OTHER ADVICE AND WORDS OF WISDOM

PREFACE

THE CHAPTERS AHEAD CONTAIN MY OPINIONS and thoughts from nearly twenty years spent crawling over, under, around, and through houses. I will talk about everything—from breaking bottles of wine in houses to what to do if you're attacked by a raccoon in a crawlspace to what type of buyer is most likely to sue you. I will also share many of my pitfalls, in hopes that you can learn something from them. I know I sure have.

This book is geared toward *concepts* related to being a successful home inspector, and does not get too far into the technical aspects of performing inspections (though you will definitely get some). I'm almost fifty and am finally getting this book done—maybe in another fifty years I'll get to a book about the technical aspects of performing an inspection. On second thought—by that time, everything technical in houses will have changed, and I will probably have forgotten everything I currently know anyway. Let's just leave it with what I have here.

You will probably disagree with some of the opinions I express in this book—that's okay. If you agreed with everything I wrote there would undoubtedly be something wrong with both of us. One of the beauties of being a home inspector is there are many different ways to approach the business, and you can run yours as you see fit.

I've often told other inspectors there are only two things that matter with respect to running a home inspection business. The first is that your phone is ringing with new business. The second is that that your phone isn't ringing with angry past clients and their lawyers. As long as those two things are happening, you are doing it right.

Away we go.

INTRODUCTION

SO YOU WANT BE A SUCCESSFUL HOME inspector? Well, I have some great news! Being a successful home inspector is an amazing career. The money is good bordering on great. You rarely have to travel away from your family. You can largely set your own schedule. And lastly, the work is really enjoyable.

The greatest part to being a successful home inspector is that no training is needed and it's so easy a child could do it. Also, inspectors assume no liability for mistakes they make, and there is an endless supply of work, regardless of the time of year or the state of the economy. It's really basic stuff that anyone who has worked on houses before can do.

Okay, time for some honesty. As you might have guessed the first paragraph is totally true and the second is totally false.

Realistically, it is easy in most areas to *become* a home inspector. Many states still don't have licensing for the job, so all you really need is a truck, a dog, and a flashlight. But being a *successful* home inspector is a different story. I'm sure many of you have talked with friends or family over the years who think they should be a home inspector because they know houses. Some are probably even contractors. I'm sure anyone who has hired contractors will readily admit they aren't all created equal and they make a lot of mistakes. "Don't worry, the seller is a contractor" is a favorite statement I've heard over the years. I'll let your imagination run wild with how that story plays out (hint: it doesn't usually end with me getting home early).

Honestly, there is a lot more to being a successful home inspector than people realize. It takes time, experience, good communication skills, good computer skills, good marketing skills, a good lawyer (often), and, most importantly, the right mindset and the desire to succeed. A wife with a steady job and a good health plan doesn't hurt either. If you're thinking of being a home inspector for some extra beer money or just something fun to do to kill time in your retirement, my advice would be to forget it. A couple decades ago that may have worked, but the profession has evolved a lot. Today it is chock-full of professionals with a deep skill set and a business model that works. Unless you take it seriously and really do your homework, the best you can hope for is barely hanging on as just a home inspector. Not a *successful* home inspector.

WHAT IS A SUCCESSFUL HOME INSPECTOR?

Before I go on throughout this book about being a

successful home inspector, I need to take a minute and define just exactly what that is.

First, let's look at what being "successful" means to me. I was once sitting in a pub finishing up one of my inspection reports for the day, when I overheard two guys sitting next to me at the bar talking about what being "successful" was. As they debated back and forth, I quickly tuned them out and began to ponder the question for myself. Having just had a birthday where I struggled to come up with a wish while staring at the cake, I realized that, as far as I'm concerned, success is being thankful for the things you have rather than wishing for things you want, while looking at the ever-increasing bonfire atop your birthday cake each year. After all, how can you ever be content with what you've accomplished if you are only hung up on things you haven't done yet?

Being successful in any career means you enjoy what you do. You don't count down the minutes away from work for fear of going back to the office. You don't dread the end of a vacation. You don't see work as a burden. I honestly love what I do most days, and when I'm away from work for too long I start to get anxious. I'd never call myself a workaholic, but there is definitely part of me that is addicted to going to work each day.

Now that I've defined what success means to me, let's take it a step further and talk about what being a successful home inspector means. A successful home inspector generally has as much work as they'd like, and makes a good living doing inspections. It's not uncommon for a good home inspector to make six figures or more. Some owners of multiple inspection companies can make substantially more. Keep in mind there are booms and busts in the real estate market, and a fair amount of seasonality inherent to this business. Even the best inspector will have to wonder from time to time if his phone is working. But aside from the natural ups and downs, a good home inspector should be able to stay busy without having to take side work or have other jobs.

And the work doesn't come overnight once you decide to go into business. In an average market it should take an inspector somewhere around three years to build up a referral base and learn what the business is about. If it's been much longer than that and you are still struggling for work I promise there will be parts of this book that can help you. Of course, factor in the market conditions in your area and the time of year before you either panic or decide you've arrived. If your area is booming, it's June, and you get three jobs in a week because everyone else in town is booked out two weeks, don't get too excited. Conversely, if it's a bad market the week before Christmas, don't get discouraged if you have a few days off.

A successful home inspector has a healthy respect for what is on the line at an inspection, but has no fear and does not second-guess their ability. If someone asks me a question that I don't know the answer to, I'm almost excited to tell them that. Of course, the next words from my mouth are that I will find out—if it came up and I don't know it, I need to know it. The point is that a successful home inspector becomes truly *immersed* in the industry, and has a desire to learn and succeed that is greater than that of their competition. A successful home inspector never sees inspecting as just "a job."

I had the perfect storm of circumstances going into this career, and I believe that helped me succeed. I had a newborn baby, a new house to pay for, no college degree, and very little to fall back on if this inspector thing didn't work out. I had tried and given up on (or failed at) numerous other jobs—and this was the best "light at the end of the tunnel" I could find. I hoped it would take me away from waiting tables and other jobs with no real future. I often liken my inspection career to the set of questions that go something like this: "How fast can you run? How fast can you run with a pack of dogs behind you?" I really had no choice but to make this work—and thankfully it did.

One last note on this whole "success" thing: Being a home inspector is rewarding to me, helpful for my clients, and serves a very important need in our society. Being able to help people understand what they are buying and how their house works can be really fun. But being a home inspector isn't life-changing in a grand sense. I'm the first to admit there are so many other people doing great things that make the world a better place. There are police officers putting their lives on the line every day, teachers shaping kids' futures, and doctors coming up with medical breakthroughs. Looking at it that way, being a home inspector is kind of a "ho hum" thing. Many days, as I'm running around sticking my electrical tester into outlet after outlet, I have to think there was something more meaningful I could have done with my life. But I am content.

PART ONE

GETTING STARTED AND RUNNING YOUR BUSINESS

CHAPTER ONE

UNDERSTANDING WHAT YOUR CLIENT NEEDS FROM YOU

TO BE A SUCCESSFUL HOME INSPECTOR YOU must understand why you are called to a house in the first place, and the sequence of events leading up to your inspection. By the time you get to a property your clients have usually spent weeks, months, or even years planning, saving, and shopping for a house. This is one of the biggest moments of their life. Don't ever forget that. You may do this every day but always remember how much any one house matters to any one client. Treat the process with respect. You'll meet arrogant mansion-buying millionaires who won't even bother to talk to you, and you'll meet struggling single mothers who will hang on every word you utter for the next ten years.

To a large extent, a house is a house and a problem is a problem—but to be a successful home inspector you must really put yourself in your buyer's shoes and give them the advice that they need. Of course, in a perfect world, everything you find wrong with a house will be repaired by licensed contractors and the buyer will live happily ever after. But have you looked around lately? It's not a perfect world out there. (Many would say far from it.) To really help your buyers you must prioritize and put into perspective what they are buying, and what a house needs *now*—versus what it will need in the future.

Some argue a downside to prioritizing repairs is that once you start assigning a magnitude to problems and offering your opinion, you take on liability for the items that you don't emphasize. Let me clarify that when I talk about helping buyers understand what's most important I'm talking mainly about verbal communication onsite at the house. When everything gets written in the report, it is presented in a more neutral way, without an opinion as to what is the most important. Your X is broken, and you should do Y about it—because if you don't, Z will happen. As for taking on excess liability by giving clients your input and advice, the good far outweighs the bad. Why should I deprive 999 good clients of helpful information, just to try and protect myself against the single nutcase who is probably going to sue me anyway? There is no good reason. The folks I truly give my best information to (at least the ones who don't sue me) remember that, and call me back for another inspection at some point. Even better, they tell their family, co-workers, and friends to use me too. Don't let the single nutcase take all that work away from you. And most importantly don't let that single nutcase ruin a good inspection for everyone else.

So, to truly help your buyer, take a step back after the inspection and run through the house a system at a time.

Here's an example of how one of my inspections might finish up on a twenty-year-old house:

"Well, that's pretty much it. Overall, I'd say it's a good-looking place and the things I found are more age-related than actual defects with the house. On the good side, the roof has just been replaced—as you'd expect on a twenty-year-old house. The siding is in overall good condition but will need paint in the next few years. And remember to get the trees and bushes trimmed away, as we talked about. The HVAC system is original and nearing the end of its life expectancy—so plan for replacing that equipment at some point. In the meantime, I'd check with the seller regarding service history, and have a heating contractor out if it's been more than a year or two. The other two mechanical systems, plumbing and electrical, checked out well, which is to be expected on a house of this age. The main things I look for are homeowners doing modifications or repairs, and really just testing everything out to be sure the systems are functioning as intended.

"On the inside, things looked okay. But there are quite a few smaller items. Things like doors that don't latch closed, a couple windows with failed thermal pane seals, and a garbage disposal that is cracked and leaks. Most importantly, be sure to update the smoke detectors and install carbon monoxide detectors. Also, updating the GFI protection to meet the current standard would be pretty high on my to-do list, due to the relatively low expense and increased safety. Hopefully, I've mentioned everything that will be on the report—but I often pull things out of my notes later so be sure to read over the report carefully and call or email anytime you have any questions."

I've rattled off a spiel like that a few thousand times over the years, and when I'm done I can see the appreciation on my clients' faces.

You can technically be the best inspector in the world, but if you can't or don't communicate what you see in a manner that people can understand and use, it doesn't matter. To be a successful home inspector you must understand where your client is coming from and what they need from you.

CHAPTER TWO

KEEPING A LICENSE, ANSWERING THE PHONE, AND OTHER BIG-KID THINGS

UNLIKE EVERY OTHER CHAPTER OF THIS BOOK, I really hope this one doesn't teach you anything. Running any business takes a certain amount of organization, planning, and execution. Keep in mind, if you do all of these things perfectly, the best you can hope for is about a 50 percent chance that you will fail in the first year and about a 75 percent chance you won't make it five years. So you'd better not just do these basic things—you'd better do them perfectly and effortlessly.

Nothing should get your attention like a letter from whatever organization holds your license. For me here in Oregon I am licensed by the Construction Contractor's Board. The state seal of Oregon is very distinct, and nothing makes me drop my electric bill faster than seeing mail with that seal. Anything related to licensing, taxes, or the government in general needs to go to the top of your list. You can be late on some bills in life, but don't mess around with the government. This advice is twenty times as important if you have employees and are late sending in tax withholdings. Essentially, you are taking money from your employees and you need to send that money to the government *immediately*. I actually get really nervous in the short time between when I pay my employees their net pay and I send the government the taxes I withheld. When it comes to owing the government money, there are all kinds of analogies about mob bosses and fingers being chopped off. Just pay your taxes first and everything else second.

As for your license and other government stuff—a home inspector with a lapsed license is about as useful as a submarine with a screen door. Beyond that, insurance, bonds, continuing education, updating addresses, keeping accurate financial books, and things of this sort are a no-brainer. If you aren't 1000 percent sure you can pull all this off, go get a job helping people put bolts in little bags at the local hardware store, because you are not ready to own a home-inspection business. Between the technical knowledge and the nuances involved in running a home inspection company, if you don't have basic business operation down, you are doomed.

Answer your phone, and answer emails day and night. This is particularly important when you are new. You *must* answer the call quicker than the other two guys, whose names the agent gave out. I get a lot of work being responsive at off hours. Most prospective home buyers work nine-to-five jobs. These people start making decisions about accepting offers when they get off work. Getting an offer to buy a house accepted starts the clock ticking on a very anxious timeline. The first call is often to the home inspector. Once you've been in the business for years and have a loyal following, you can safely return calls the next morning. But when you are trying to grow your business and gain market share you must find a way to be receptive to scheduling at odd hours.

As a final piece of advice on phone etiquette: no screaming kids in the background. I know having Mom pull double duty watching junior and answering the phone for your company seems appealing, but don't do it. There is nothing that makes a business come across like a low-budget operation more than forcing buyers to yell the address for their prospective million-dollar home over the high-pitched screams of a two-year-old who is late for a nap.

Drive a decent vehicle. It may be time to retire the old "Trusty Rusty" 1996 Chevy pickup that has been with you through so much. The off-color quarter-panel from that epic camping trip with your buddies, and the jacked-up bumper from when you helped pull your brother from a ditch are nice memories—but unfortunately, the buyer and agent weren't there, and don't understand. I know Mom always said not to judge a book by its cover. Sorry, but the agent and buyer had a different mom. They do judge you by your cover, the instant you pull up. I'm not saying you need a new top-of-the-line truck. Honestly, vehicles are amazingly affordable these days. You can get a lot of truck for your money. For five grand, you should be able to get a plenty presentable ride—in which all of the body panels are present and matching. If you can't afford (or borrow) five grand for a vehicle to start a business, I don't have good news for your chances as a successful home inspector. I know the world shouldn't work this way, and if I were king it wouldn't. But I'm not and it does.

CHAPTER THREE

CUSTOMER SERVICE ABOVE ALL ELSE

BENDING OVER BACKWARDS TO TAKE CARE OF customers is the wave of the future. You'd better jump on board or you will be left behind.

This concept started slowly a couple decades ago, and has been gaining steam ever since. There is a reason stores like Costco and Cabella's are wildly successful. Their employees are always helpful, and the stores take returned items back with no questions asked. This gives shoppers great assurance when purchasing something—and the word of mouth sharing the experience spreads like wildfire.

What can we take from this and apply to our business as home inspectors? There is a lot, actually. Contractors and others involved in the construction business are notorious for being unreliable, hard to deal with, poor communicators, and just generally difficult. As of this writing, a good portion of the home inspection industry is still this way. Honestly, I feel lucky that my competition makes it so easy for me to stand out by taking care of my clients before, during, and after an inspection.

To give a little background on how I arrived at this business model, let me tell you about what I did before getting into home inspections. For fifteen years, starting in my late teens, I worked for a corporate chain of restaurants. As a sixteen-year-old kid with a marginally poor attitude I started washing dishes. Next came bussing tables for a while. I then got into the kitchen, where I worked as a prep cook and then worked my way through all facets of the cooking line: fry cook, pantry, sauté cook, broiler, and lead. After being dissatisfied with a cook's meager hourly rate, I moved out into the front of the house, where I waited tables for over ten years, and did a bit of bartending. It still strikes me as unjust that, including tips, a waiter makes far more than the person cooking the food—but that's just the way it is.

The company that I worked for had a couple dozen restaurants scattered over a few states, and an extremely organized and efficient business model. The customer service standards that I was held to seemed a bit over the top at first, but as I matured and started to understand things, I realized they had struck gold. If someone didn't like what they ordered, the meal was replaced with something else. If they felt the food took too long, most times the manager would comp it. If the customer disliked the attitude of some employee, the manager would apologize profusely, and the offending employee would be reprimanded. The company's fundamental belief was that the customer is *always* right. (This saying was often echoed by the company's chairman of the board, Bob Farrell, of

Farrell's Ice Cream Parlors—he wrote a great book called *Give 'Em the Pickle*.)

What I've realized over the years doing inspections is that if I treat my customers the same way as I was encouraged to in the restaurant, it will come back to me tenfold. Anytime the phone rings with a problem I look at it as an opportunity to turn a bad situation into something good. Of course, it doesn't turn out that way every time—but most of the time it does. When one of my good agents calls and tells me the bank is holding up closing because they need proof the fallen fan duct in the attic has been re-hung, I don't tell them to call my office and make an appointment. I get someone onsite ASAP and get it taken care of. You'd be amazed how powerfully that kind of positive attitude resonates within an industry, especially by word of mouth.

One of the best agents and nicest people I've met since becoming an inspector is a guy I picked up by standing up and dealing with a problem. I did an inspection on a mostly vacant house with dozens of electric heaters. There were baseboards, ceiling radiant heaters, and Cadet in-wall types scattered everywhere in the house. The potential buyer had what seemed like twenty kids running around, turning things on and off. Apparently one of the heaters got left on. Well, my client didn't buy the house—and the lady looking after the place didn't realize a heater had been left on until she got the electric bill a few weeks later. It was about $150 more expensive than usual. When the listing agent called me, I remembered the house, and, without putting up a fight, told him I'd gladly pay for it. I'm reasonably sure one of the kids turned on a heater, but I'm not above admitting it could have been me. Honestly, it really doesn't matter. At the minute that phone call came in I could have denied responsibility and left the agent thinking I was a jerk, and that my company was just like all the others. But for $150 I bought the best advertising available. After this situation that listing agent started referring clients to me—I've done well over $10,000 of inspections for his people over the years. It constantly amazes me that my competition will spend thousands of dollars on advertising—then do everything they can to make sure clients never come back.

I like to hang out on internet message boards with other inspectors, and have learned a lot over the years. I remember once the topic of printing and mailing copies of a report came up. For you kids out there—before the days of email, it was standard to either print a report onsite or back at your office, and mail it to the client (thankfully this is no longer the case). Anyway, I remember several of the posters on the message board saying that to mail out a hard copy, they charged twenty-five cents per page, plus postage. Are you kidding me? Someone just gave you $400 for a few hours of your time and now you want $5.30 to mail them a report? Wow, it must be nice to have so much extra business that you don't have to worry about pleasing the clients you already have.

Here's another lesson that, if you are already a home inspector, you have undoubtedly encountered. (If you aren't yet an inspector, plan for things like this.) Let's say it's your afternoon inspection, and upon starting, the agent informs you the buyer is "really busy" and will show up for a walk-through at four. Fair enough—you should be ready by then. But when four rolls around, no buyer is in sight. At some point (usually at least thirty minutes after expected) the buyer arrives—usually with a fresh cup of coffee she figured she had time for. She insists on a very detailed explanation of *every* problem in the house, beating each little issue to death over and over. "How much is this going to cost? Who is going to fix it? When will it get done? Can you come back and be sure they did it right?" And that's just trying to explain that a smoke detector needs to be replaced—wait until you get to the problems that require some thought!

Just when you think you're finished, the husband shows up and expects the same treatment you just gave his wife (are couples not capable of communicating anymore?). By this point, I'm often extremely annoyed, and can't believe how inconsiderate and self-absorbed some people are. Ironically, these folks are buying a house *much* nicer than mine. They will often expect you to get the ladder out of your truck to take them back up to the attic, or walk to the depths of the crawlspace to show them a hanging piece of underfloor insulation. The fact that you were in these very same places two hours ago and had finished your job never enters their mind. At moments like these I honestly have to pretend to be happy and in a good mood—because if my true feelings were to show I'm sure a slew of one-star online ratings would follow.

One piece of good news in cases like this is that agents will usually realize what is going on and be very thankful for how you handled the situation. After all, they were ready to be done an hour ago as well. As you pile yourself in your truck and head for home, having missed your kid's soccer practice, try to remember the good days and the good buyers. Thankfully, there are more of those than the bad ones.

To be a successful home inspector, develop the mentality that the customer is always right, and do anything you can to keep them happy. Look at problems and difficult people as opportunities to show that you run a successful business, and you are serious about taking care of your clients. I promise: the money and time you spend will yield more returns than any other advertising you do.

CHAPTER FOUR

REFERRALS FROM REAL ESTATE AGENTS

ABOUT ONCE A YEAR I GET A CALL FROM A SKEP-tical homebuyer who got my name from their real estate agent. They are skeptical because they think their agent and I are somehow in cahoots and I'm going to gloss over problems in their prospective house just so the agent can close the sale and get a commission. While I understand their concern, anyone who really knows how the relationship between a real estate agent and a home inspector works will understand that the skeptical view isn't a successful business model for either side.

Let's look at it from the agent's side first. Any good real estate agent will tell you that the first step to building a successful business is getting clients and referrals to friends and family of past clients. Yes, there are a number of homebuyers who will just wander in off the street or call them out of the blue. But successful agents make a living on "strings" of business—family member after family member, co-worker referrals, friends of friends, and so on. Referrals are so important to real estate agents that most have some blurb on the back of their business card or in the signature of their email that says something like, "I'm never too busy for your referrals." Nothing is more important to an agent than having a buyer walk away from a transaction with a positive feeling.

As an aside, today a consumer's feeling walking away from any purchase is more important than ever. With the popularity of social media, and a buyer's power to spread word of a bad experience all over the world with a few keystrokes, companies and anyone selling anything know that the key to success is happy customers.

With this in mind, we have to ask: how would it benefit any real estate agent to get a home inspector to "gloss over" problems in a house, just to make a sale? I'm hard-pressed to come up with a reason.

Now, with all the real estate agent's motives aside, let's consider whether there's a benefit for a conspiratorial home inspector. About the only things I can come up with would be an increased number of lawsuits, an abundance of phone calls from angry clients, a bad reputation in the industry, and a catastrophic lack of work. I know—you're thinking it all sounds too good to be true.

After talking to a skeptical buyer I will congratulate them on having enough awareness to question the process, and thank them for at least taking the time to call me. I'm sure for everyone who calls me there are ten or more who got my name and threw it in the trash, since it came from their agent. I don't really blame them. Honestly, I'd probably do the same thing if I didn't understand the way the process worked. In the end, I perform inspections for most of the people gracious enough to call me, and willing enough to let me explain. I actually remember many of them, and they have turned out to be some of the best folks to work for.

To be a successful home inspector, you must be comfortable with the relationship between real estate agents and inspectors. The reality is, agents are the ones most likely to send business your way. If you fight that, your career is going to be an uphill battle.

CHAPTER FIVE

A GREAT
MODEL
FOR GROWING
YOUR BUSINESS

THERE'S NOTHING LIKE FREE ADVERTISING, right? But how about getting paid while you advertise?

Of course the second option is better. And that's exactly what happens when a past buyer or a buyer you otherwise know lines you up for an inspection. In almost every case that buyer is going to have an agent, who will probably want to refer home inspectors to their clients. Getting in front of agents is, without a doubt, the best way to grow your business and gain future referrals. Many grumpy home inspectors cast stones at those of us who try for realtor referrals. I'll have much more to say on this later, but for now, let's just focus on the dynamic of getting in front of new agents.

Once you get that appointment and get in front of the agent, you really don't do anything different than you ever do at an inspection. It's kind of like a first date. If you make a fool out of yourself trying too hard you're very unlikely to get a second date. Rather, you just do your normal routine and let your work speak for itself. I promise the agent is already somewhat curious about you, since their buyer thought enough of you to insist on calling you. Most likely the agent was trying to push "their guy" on the buyer, but the buyer fought that off.

The most I'll do for marketing to a new agent at an inspection is to exchange business cards. Depending on the situation I might make some conversation and talk about people I might know in their office. I try really hard to not try too hard. I find that actually goes much further than pursuing them more forcefully. Once I've worked with them, they are in my database, and I might send out an occasional informational email. I tend to keep these pretty infrequent so as not to be a pest. Other than that, I just let my work and fate take over.

I've heard amazing stories from some of my regular agents about the lengths inspectors will go to in trying to steal them away from me, in cases where their buyer insists on choosing the inspector. I'm always excited when I see an agent and they tell me they worked with someone else. I love to hear about the experience. Most times the things the agent reports make me feel even better about what I'm doing. Of course, I'm certain I've lost clients to other inspectors as well. I don't for an instant think there aren't better inspectors out there, and I recognize that some agents just want someone with a different personality. The trick is to be gaining more than you are losing.

This leads me to the related topic of agent referrals. These are great, but agents come and go like the wind. Some of the best ones I've worked with over the years have just seemed to disappear one day, for whatever reason. They'll often resurface sporadically, but it's clear they found someone else to work with. I rarely ask why, and in most cases I have no idea what happened to make them go elsewhere. To me this is just a natural part of the business and something I don't spend much time worrying about. I'm not going to change the way I do inspections to appease an agent so it really does me no good to worry about why they might have left.

Real estate agents are a great referral source, and you should definitely be aware of them. Just don't get caught up trying to figure them out, or court them.

CHAPTER SIX

UNIFORMS, VEHICLE SIGNAGE, AND BRANDING

I LEAVE PLENTY OF MONEY "ON THE TABLE," AS the saying goes. I probably just flat out give away a few dozen inspections a year to my competition. I have no signage on my company vehicles, my employees wear no logoed uniforms and, other than marketing, email, and printed materials, you won't find my logo or company name anywhere.

Why? I have a variety of reasons. During my time hanging out on message boards I saw this topic go around somewhat frequently. The two schools of thought were to either brand and market everything, or hold yourself in a higher regard and act like your accountant or attorney. I generally agree with the latter but, of course, I also realize attorneys go to years of school, and sit at nice desks overlooking the city rather than in crawlspaces looking at dead cats. So to some extent, it is an inaccurate comparison. When was the last time you saw a car on the road with a vehicle wrap that said, "Harris, Bowker, and

Cunningham, Attorneys at Law?" I generally like the idea of wearing plain clothing and not jumping in with all the glitz and branding. I like to think I am a consulting firm, not Zippy's Dry Cleaning Service.

I've been accused of being a skeptic, so some will disagree with my opinion on this. It just seems like the whole world has turned into a franchise of some sort. My immediate feeling when I see some poor guy in a logo-riddled truck and a bright-green button-down collared shirt is that he's making fifteen dollars an hour, and the genius who started the company is making millions. Sometimes this is okay, but I don't want people thinking my company is staffed by fifteen-dollar-an-hour guys. People are paying for more than that, and I want to deliver a quality impression right from the start.

Don't get me wrong. It's not as though I show up in a tank top and surfer shorts. My company's dress code is generally just plain clothes. No large brands or cute sayings. Definitely nothing political or religious. I like blending in with things in the world. In traffic, while parked in neighborhoods, whatever. Sometimes, I don't want to be a home inspector—I just want to be a guy going to Home Depot on a Saturday morning. Folks in business trucks get hassled constantly. I can imagine a home inspector at Home Depot is a particularly easy target. And, as a final benefit to leaving logos off my company trucks, I don't have to field phone calls from angry folks who one of my inspectors cut off in traffic and gave the finger to.

I understand my opinion flies in the face of all the "rah, rah, rah" marketing folks, and I guess I'll have to accept the fact that I'm leaving money on the table. Like everything else with this business, research all the angles and opinions, and make the decisions that make sense for you. Remember earlier, when I said there were only two things necessary for being a successful home inspector? Number one is that your phone is ringing with people requesting your service. Number two is that your phone *isn't* ringing with angry clients and their attorneys.

As long as both of those are happening, you are doing just fine.

CHAPTER SEVEN

MARKETING

THIS IS A TOUGH ONE, BECAUSE, TO BE HONEST, I can't tell you anywhere near all of the things I do, since to some extent you are my competition. I suppose I could have just omitted this section, but marketing is such a massive part of what you must do to be a successful home inspector that I didn't think it would be fair.

Marketing isn't just a website, or a magazine ad, or word of mouth, or vehicle signage, or a newsletter, or follow-up emails asking for reviews, or a blog. It's all of these things and more. Like any business, home inspection companies have roughly a 50 percent failure rate. I was lucky enough to pay a small fortune for an existing company, so in a way I bought a client base. This eliminated the need for me to really be good at marketing. I'm personally not that great at marketing but I do spend a good amount of money maintaining my company and expanding my market share. In my area, the lion's share of the jobs go to the well-established companies. There have been some newer folks eating up some market share, and without fail they are all fantastic at marketing. I suppose they have to be good inspectors too, but a lot of them I don't really know. From a strictly business-volume standpoint, a bad inspector with a well-implemented marking strategy will outlast a good inspector with bad marketing. No one ever said the world was fair.

There you have it—the shortest section in the book, but probably the most valuable information for becoming a successful home inspector.

PART TWO

STAYING OUT OF LEGAL TROUBLE

CHAPTER EIGHT

LIABILITY, LAWSUITS, AND GETTING SUED

HERE'S ONE I COULD WRITE AN ENTIRE BOOK about. As you look into becoming a home inspector, you have likely been made aware that you take on some liability when you perform an inspection. But if you haven't been hit with this fact of life, let me be the first to explain it. You take on a *huge* amount of liability performing an inspection for money.

Let's start with the definition of the word "liability."

According to the Google dictionary, liability is "the state of being responsible for something, especially by law."

So what does that mean, and what does it have to do with a home inspection? It usually goes something like this: John Homebuyer puts an offer on a house for a given price. Then he hires you to go through the house and find problems. Next, John re-negotiates with the seller to either fix things you found or give him a price reduction. When John moves in after the seller is gone and finds more things that cost money to fix, he naturally wants to blame you.

I say "naturally" because this has become the norm in our society. There is no such thing as something just happening. No matter what it is, someone is at fault, and they should pay the bill. Roofs don't just wear out, and pipes don't just leak. Someone is responsible—and as the last one to look at the place, you are the guy with the finger pointed at you. If you can't handle this position, you either need to learn to accept it, or find another profession—because this pattern isn't going away soon.

The good news (if there is such a thing on this topic) is that the home inspection industry has been around for a long time, and inspectors before us have paved the way by dealing with many frivolous lawsuits—and some valid ones too. That experience shows that it is crucial to have a good contract drawn up by an attorney familiar with this industry. Also, keep in mind that liability issues in the home inspection industry are a moving target. You can't just visit an attorney when you get into the business and then forget about it. I regularly poke around online for inspection case law, and keep an ear out for things in the industry. Be sure to keep up on what's going on with your local area, and especially the people who license you (if your area has licensing). It doesn't matter at all what's happening in Texas if you live in Michigan and there are new laws being made in the latter state. From talking to inspectors around the country on message boards and at conferences, I'm constantly amazed at the differences between the various states.

I attend a national conference in Las Vegas every

couple years, and it tends to draw a lot of inspectors from nearby California. During the presentations I've heard a lot about California's laws and rules, and it's pretty interesting stuff. The last I heard is that limitation of liability clauses were not allowed in home inspection contracts, and that arbitration clauses might be next to go. What this essentially would mean is that a buyer could just drop a lawsuit on you and you'd have to go straight to court. Historically, one of the best tools a home inspector has to defend himself has been mediation and arbitration, which is much less costly than a courtroom. Also, a limitation of liability clause has been crucial in the past to put a cap on the damages an inspector can be expected to pay. The fact that all of these things could be wiped away is frankly concerning, and I hope it isn't a sign of things to come for the industry as a whole.

To be a successful home inspector, you must have a well-written contract from a good lawyer, educate yourself on how legal issues work in your area, and always keep your ears open for changes that impact you.

CHAPTER NINE

WHEN THE CALL COMES IN

AS REWARDING AND FUN AS BEING A HOME inspector can be, the phone can ring with an angry client at any time. They might be holding a broken pipe or rotted floor joist at the time. Or maybe they are holding a bid from a contractor who is there to fix something you "missed."

Nothing interrupts a nice sunny day listening to a good song on the radio like a complaint call. I've been called in the middle of a vacation, the middle of a nap, the middle of another inspection, and the middle of the night. If you have a conscience, like I do, these calls take over your thoughts for a few hours or days—or, in really bad instances, weeks, or even months. In the end, complaints and problems are just part of the business. Nine out of ten times the hype and hysteria associated with the call turn out to be misguided, or the result of a misunderstanding. In most cases the problem can be taken care of for little or no money. Regardless, I still get rattled every time "the call" comes in.

Let me give you a little context so you don't think I'm a terrible inspector who makes a lot of mistakes and gets complaint calls all the time. I began my career by working for the oldest, biggest inspection firm in Oregon. After I worked as an employee for several years, the founder retired and sold the company to myself and one other employee. We inherited several employees and hired some more, and ran what I consider to be a very successful company together for roughly thirteen years. After that, I bought out my partner and now own the company alone. At any given time there are between three and six employees performing inspections every day. This translates to roughly five times the number of inspections as a typical single inspector company, and five times the exposure to problems and callbacks as most companies. As I talk about problems arising out of past inspections, keep these statistics in mind.

So, what do you do when the call comes in? First of all, keep in mind that the person making the call has had a chance to prepare for this moment, whereas you are totally caught off guard. It's important to remain calm and try to gather as many facts as you can. Don't ever defend anything or try to make an excuse during this initial discussion. Like you've probably heard numerous times, a lot of people just want to vent and be heard. So you should listen more than talk. And *really* listen. Don't just sit dead on the other end of the phone, waiting for them to finish. Listen for key things, like what has been done. Who have they paid, and who are they about to pay? In a perfect world, this first call goes to voicemail or comes through your office, so you can be somewhat prepared. If you're unlucky and just answer the call without expecting it to be a complaint, do your best to listen and get the basic information you need to do some research. Get the client's name and address, and the date of your inspection. After listening and gathering this information, assure them you take problems very seriously and need to do some research and will get back to them. Give them a timeframe when and how they will hear back from you, and tell you you will likely want to get onsite as soon as possible to see the situation. Encourage them to hold off on any repairs until you have a chance to look at the problem.

The way you handle yourself at this critical point often sets the stage for an entire sequence of events, and it is important to be cordial and attentive, and treat the client with respect.

Okay, so you're through the initial phone call/complaint and you now have a "situation" on your hands. These complaints are an unfortunate part of this business that there is no way of avoiding.

If you have errors and omissions insurance you can call your company but they will often just want to open a claim. Of course, like any insurance, claims are bad and will likely raise your rates—and if you get too many of them, the company will drop you. This might also be a good time to read the fine print of your E&O insurance policy because many require you to immediately contact them when a potential claim arises. To me, this is a bit of a slippery slope because you don't always know when something is going to turn into a full-blown claim. Notifying your E&O carrier every time there is a hint of a problem is likely to get you canceled, so my advice is to at least investigate the alleged problem a little on your own first.

For the sake of argument, let's assume you have no insurance, or are just going to try your best to work out the situation. After the initial call from the client, you have enough information to pull the report, look at what you wrote, and to try to remember what you can from the inspection.

Now that the initial shock of a complaint has passed, I can give you some encouraging news. Over the last sixteen years we've had roughly one hundred and fifty complaint calls. Seventy-five went away without an inspector having to go onsite. Forty went away after an inspector went onsite, and talked to people and explained things. Fifteen went away with a refund of the fee, or something close to that. Three required a reimbursement of something above the inspection fee, but less than $5000. Two were the ones you lay awake at night worrying about.

Really, this is good news. When you field a complaint call there is roughly a two percent chance it's going to be a huge problem. Keep in mind this is just based on my experiences. If you are the one in the "huge problem" group you aren't going to care that I got off cheaply ninety-eight times. But when getting complaints, it's generally good to remember that most of the time you're not going to lose your business, or cost your E&O carrier six figures.

CHAPTER TEN

PUTTING
OUT FIRES

SO, YOU HAVE A COMPLAINT FROM A PAST inspection. You've been contacted, done some research and hopefully gotten onsite to see just how bad (or not) your liability is. Now it's decision and negotiation time. As a general rule, bring your checkbook and a release-of-claim form to any onsite visit. Often, writing a check early in the game is the best way to shut a situation down. It can hurt your pride and tug at many other emotions, but I can tell you from personal experience that those feelings fade more quickly than a nagging, lingering situation. It is also often less costly than paying for attorneys, arbitrators, or mediators.

As a general rule, I'm happy to just refund an unhappy customer's money. After all, I really don't have a product cost. It's just my (or my employees') time. Many folks are just happy to take their money back and move on. It's just human nature. You paid for something, and it is not what you expected, and you are happy with a refund. There is also that "limitation of liability" clause in the contract they signed. Most people have enough pride to abide by what they signed, or understand it will be an uphill battle coming after you. In case they don't realize the latter, I will often point it out in the contract they signed. Some basic research also backs us up on this. If you google "Can I sue my home inspector?" you'll discover how hard that is to do.

You might notice that I'm really not getting too far into whether or not I missed something or have any fault. Early on in the game these things are somewhat irrelevant. Sure, if you get to arbitration or go to court these will be huge factors, but early on I never defend too much. I just try to satisfy an unhappy customer. Of course, I talk to them about the problem, and do some explaining, but until we are talking about money well above the original fee, I just don't see much benefit in taking a defensive stance.

To me, the complaints fall into two categories. First, the cheap and easy ones, where everyone is generally nice, and folks take their money back, and we all move on. Second, the long, drawn-out ugly ones. Again, these are rare, but they do come up. With the bad ones I will start bringing up things like the contract they signed, the state standards of practice, and limiting factors at the time of the inspection (stored items that prevented me from finding something, etc.). I'll often have to dispute the opinion of an overzealous contractor who has dollar signs in his eyes. These guys are great—they think they're going to remodel the entire house and send me the bill. And they are more than willing to tell my client that I should pay. "The contractor said you should have seen the rotted floor that he found when he tore up the tile," clients will inform me. "He can't believe you missed it."

There's a funny trend in the complaints we've had over the years. The people with big problems that our inspector honestly missed are the most understanding and easy to work with. The cases that turn into nightmares are generally frivolous and without merit. The most egregious example of this was the CEO of a mid-size bank. In researching the case, I discovered that he made $35 *million* (yes, you read that right) the previous year. Yet he came after us for a rotted deck that had been totally concealed. He messed around through two mediations, and finally took $2000. I suppose if the guy sleeps better on his goose-down bed with my $2000 in his pocket, I'll just have to accept it. One thing I can promise you: this business is entertaining from time to time.

In the end, my best advice is to be ready for complaints and get a thick skin. Complaints are part of the business. Sure, being a technically sound inspector helps, but there is nothing you can do to completely eliminate the chances of having liability issues. It's just a fact of life and a product of the society we live in. If you are in this business for any length of time you will get call-backs and complaints, and will probably be sued a time or two. Plan for it going in, and hopefully you can minimize how much it consumes you when it happens.

CHAPTER ELEVEN

WHO YOUR CLIENT IS AND HOW LIKELY THEY ARE TO SUE YOU

GOD BLESS DOCTORS. THEY LITERALLY KEEP THE world getting out of bed each day. Without doctors our life expectancy would probably be half of what it is. By and large, doctors are a great bunch. They've given the majority of their lives to their careers, spending countless hours in classrooms and years getting stepped on as a resident—all before making any real money.

But aside from all these great things, they are also statistically the most likely to sue their home inspector. Why? I don't think anyone knows for sure, but I have a few ideas.

Doctors themselves often get sued. Their insurance rates are astronomical. I promise you there isn't a move any doctor makes these days that isn't accompanied by the fear of a lawsuit. Doctors have a lot of money and buy really nice houses. A lot of doctors don't know a lot about houses, because they have spent their lives learning about bodies, not roofs. Add all these things up, and you have the perfect storm: someone who is no stranger to liability, who has a lot of money, who knows little about houses, and who probably has a lawyer buddy from college on speed dial.

As a home inspector, you will meet all types of people, from all walks of life. Some will be rich. Some will be *really* rich. Some won't. Statistically speaking, the ones with money are much more likely to come after you. Most people think money should buy perfection. Money should insulate you from problems and enable you to see the unforeseen. Of course, this isn't the way the world works. But try explaining that to someone with a lot of money.

The other side of the coin is the working class. I have a much easier time relating to these folks. In my experience they just tend to be easier to deal with. Folks who work hard for a living know that they have to look out for themselves, and that there are no guarantees in life. They didn't get some advanced degree from a nice school, and they aren't done thinking and working for themselves. Working folks accept imperfections—and are okay with them, within reason.

So, how does all this sociological rambling relate to home inspections? Basically, you'd better be on top of your game and extra picky with that doctor's nice house in the hills. A little bit of sloping in a floor due to some typical settlement wouldn't even make the radar for most blue-collar workers. To them, unless the floor is going to collapse they really don't care. But you'd sure better point it out to the doctor's wife, who will move to the Marriott, claiming she is too distraught to set foot in the new house if her China hutch is 1/64" off level.

So far, I've pointed out two extreme ends of the

spectrum—most of your clients will fit somewhere in between. The main point I'm trying to make is that it's important to be aware of the person you are working for. Your job involves more than just the house you are inspecting. The longer I do this, the less surprised I am at the people who call back and complain. When I sense someone is just a bit picker than normal, and that they may not really be hearing me, I make sure and emphasize my points on the report. I also find myself getting a gut feeling about some folks. You must always remember to protect yourself from lawsuits—being aware of those most likely to file them is critical.

One last note—I often talk to other inspectors about liability and different types of clients. I find many inspectors are afraid of lawyers like I'm afraid of doctors. I have a completely opposite outlook. I love lawyers—when they are my client, that is. When the fantastic, polished contract that I furnished them is signed, I feel like I could do just about anything and be okay. In my state of Oregon, limitation of liability clauses are still allowed and reportedly upheld. The wording is in huge bold letters in my contract, and once my attorney client signs it, I'd sure like to see him or her try to get out of it. What's their argument going to be? They didn't understand what they were signing? I don't even need an attorney of my own to blow holes through that one.

Honestly, I'm most afraid of the single mother first-time homebuyer. Mediators, arbitrators, and judges have a hard time looking them in the eye and deciding against them. I'd much rather take my chances with some hotshot lawyer.

CHAPTER TWELVE

DEFENSIVE WRITING

THE REPORT YOU WRITE IS PERMANENT. IT'S LIKE in the movies when the cops throw the guy on the ground and say something about how "everything you say can and will be used against you." Nothing could be truer in terms of being a home inspector. The only difference is that in this case it's "everything you write can and will be used against you."

Yes, every word you type could ruin you under the right (or wrong) circumstances. To be a successful home inspector you must always remember, as you're writing, to protect yourself above all else. Nothing you find should be construed as the sole cause of anything, it should always be called a "contributing factor." Nothing will eliminate anything, it should always be suggested that it will "help reduce" it. Speaking in definite terms is something you have to largely avoid in reporting. It's really too bad it has to be this way, but in our society today it is a reality.

Once you get a good number of inspections under

your belt, and you start to feel comfortable with writing reports, you will begin to look at house problems differently. You will see not just the problem but also how to communicate verbally to the client, and how to write the report in a way that serves your client's best interests and also protects you. To serve your client's best interests, you must give them clear instructions as to what to do. By a happy coincidence, this usually protects you as well.

Here's an example: consider a nineteen-year-old three-tab roof shingle with a typical service life of twenty years. The shingles have a heavy amount of mineral loss, there is heavy cracking on the surface of the tabs, the metal roof flashings are worn, and the roof is at the end of its life ... but it is not leaking anywhere. After describing all of the roof's unfavorable characteristics (complete with pictures, of course), you recommend "further roofing contractor evaluation to determine the remaining useful life and eventual replacement cost of the roof." This serves the client, as you give plenty of evidence to help negotiate with the seller, who is undoubtedly pointing to the fact that the roof is not leaking. And of equal importance is the fact that you protect yourself. That roof should be off your plate for good and you are unlikely to ever hear about it again.

Of course, you would be completely within the good graces of industry standards of practice to just report what you see on the roof and say nothing more. After all, the roof is serving its intended purpose, so there's no recommendation needed, right?

In a perfect world, I suppose that would be true— but not in the real world. This is a great example of how standards of practice are minimal, and must be exceeded at times. If you take the minimal approach, the roofing contractor who looks at the nineteen-year-old roof when it starts leaking someday is going tell the owner that you were incompetent, blind, and whole bunch of other things not fit for this book. Then the homeowner will come looking for you. If you had taken the more thorough approach, you'd be able to pull up the "roofing" section in your report, scroll down and read to the buyer just what you said. And you'd be able to ask him or her what the contractor you recommended *before* he or she bought the house said. And that would likely end the phone call.

Trust me, scenarios like this play out all the time. I honestly think only about a quarter of clients actually read the report.

Here's another important concept in protecting yourself with your report: What you don't say can cost you. Consider this write-up: "The flexible heating duct under the master bedroom has been damaged by animal activity in the crawlspace. Replacement of this section of ducting is recommended to prevent the loss of conditioned air under the house." That seems like a reasonable thing to write up, right? Possibly. Just be aware of what else you're implying here: that every other heat duct in that crawlspace is perfect and needs nothing. As long as you are okay with that, the previous write-up is just fine. Just be sure you realize what you are implying by the way you describe the things you write up. The way this heating duct example can take a sour turn is when your client negotiates a few hundred dollars from the seller, with plans to take care of the duct after he owns the house. A couple months after moving in, he calls out the contractor, who goes into the crawlspace for the repair. When the contractor moves the damaged duct, it provides access to previously inaccessible ducts, and they are also damaged. The total bill is $1500. Guess who the buyer is going to call?

The lesson here is to always have in the back of your mind just how big of a problem can be uncovered during the course of a given repair. *Warn* people! This helps the client and protects you. Both are very important.

Get out your highlighter, because this next sentence may well be the most valuable thing you ever write in a report. "The presence and extent of any concealed damage cannot be determined without removing some materials to perform the necessary repair." This is a very powerful statement in a report, and it greatly protects you. It also protects your client—it educates them and helps them realize there is no way of knowing how much money will be needed to perform the repair. And this isn't just some excuse I'm making up to try to avoid doing a thorough inspection. Many contractors—whether for roofing, bathroom remodeling, or what-have-you—put a caveat in their bids that address this exact concept. Usually, it says something about the possibility of additional costs if unexpected, unforeseen damage is found during the demolition process. In fact, many contractors put in this layer of protection even when they don't see a problem. When I write this as an inspector, I've already found some damage and am just warning that there could be more than what is visible.

The possibility of additional damage exists in virtually any repair, but the most important place to pull this out of your bag of tricks is with water intrusion. Building envelope leakage and bathrooms are the most common places to be surprised about what is found as the materials are ripped out. Of course, there are many other places surprises can occur, so always remember to warn people about what can be found as things are torn apart.

To become a successful home inspector you must always write defensively, and remember that everything written or implied can come back and be used against you. Keep that in mind with every report you generate—no matter how busy, tired, or distracted you are.

As you look at framing, remember you aren't a code inspector or an engineer. In most cases when I look at a deck or framing installation I develop an overall opinion of how and who built it. Does it look generally well-designed? Are the framing connections and fasteners well done? Is there visible sag? Does the whole deck shake when I lean on a post? I don't open the code book and look at a span table, and when there are things wrong I don't promise anyone that what I see is the complete list.

If you guarantee your clients that you will detail everything wrong with everything, you will find it takes eight hours to perform an inspection—and as a reward for all that hard work, you will likely be sued. All it takes is a contractor to come marching in and find—or claim to find—one thing you missed. Then the project turns into a much larger endeavor, the buyer claims he "never would have bought the house had he known," and his attorney will have him staying at the Hilton because he is too distraught to live in the house. No, I don't care to be part of all that—so I use a simple little write-up: "Further evaluation of all aspects of the deck is recommended to prevent a failure under a heavy load and for safety. Among the problems are the following (this is not intended to be a complete list): incorrect fasteners for the joist hangers; over-spanned, sagging framing; untreated wood posts buried in the ground and rotting; missing/incorrect hand and guard railings."

Report writing is an absolutely critical element of being a successful home inspector. Don't worry when it doesn't come easily at first. It never does, and every new inspector feels completely overwhelmed at times. Your best chance for survival is having a comprehensive set of pre-written comments to put in reports when you start, and by reading as many other inspectors' reports as possible. These things are most easily accomplished by going to work at an existing firm, but I realize that's not for everyone.

CHAPTER THIRTEEN

INSURANCE

LET'S TALK ABOUT WHAT HAPPENS WHEN YOU cause some *really* big damage in a house—like burn a house down or cause a flood. For years I had a misconception about liability insurance, versus errors and omission insurance. I just assumed everything that was damaged was covered under liability insurance. It turns out that is not correct. There is something called a "professional services liability waiver," and it is built into many liability policies. In English, this means if you cause damage while performing your professional services, there is no coverage.

For example, let's say you are testing a tub and wander off—and it overflows and floods the house. In this instance you may have no coverage under your general liability policy. Liability is only for accidents—things like kicking over a lamp into a grand piano. In order to have coverage for causing the tub to overflow, you must have errors and omissions insurance. This isn't required in every state, so you may or may not have it. But it's something to consider when deciding which insurance coverage to purchase.

Deciding what insurance coverage to have is strictly a business decision, and only you can make it. As I mentioned, many states don't require E/O insurance. But for a new inspector I would strongly encourage you to get it. During your first few years, you are more likely to miss something big and get sued. If you're an experienced inspector and you decide to run without E/O insurance, I won't say you are crazy. E/O insurance can be a bit of a lightning rod for liability. I've spent a lot of time both with and without E/O insurance, and on more than one occasion, *not* having it actually paid off. For some reason, lawyers are only interested in chasing a pot of money. Go figure.

My best advice is to go overboard educating yourself about types of insurance and your local market. You also must factor in what you have to lose. What are your assets? Are they protected? How long have you been inspecting? Everyone's situation is different, and there are things in this area that I'm not qualified to advise you on. What I am qualified to tell you is that it's a very important subject, and you need to look into it thoroughly.

CHAPTER FOURTEEN

HOME INSPECTOR FOLKLORE OR FACT?

THROUGHOUT MY TRAINING AND WORK AS A home inspector, I have heard many theories about why we should do some things and not others. Terms to avoid or put in reports, tools to avoid or use, items to avoid or inspect are all subjects about which home inspectors have various strong opinions. What follows are some highlights of home inspection folklore and fact.

Some home inspectors say not to mention the "code" word in your reports. They say if you talk about a building code in one part of the house you have just elevated the client's expectations that you are performing a code inspection on the entire house. Of course, we know that according to all standards of practice, home inspectors are not code inspectors, and are not required to inspect a house with the building code in mind. And this is for good reason. As inspectors, we are looking at houses built during the past century or more, and it just isn't realistic to expect one person to know every code or code change across every system in houses throughout that time period.

So, if home inspectors shouldn't talk about building codes, how are the codes still useful? The answer is that they are a guideline for how things should be done. Just because you don't write something up as a code violation doesn't mean you can't use building codes at all. For example, the current code for railings specifies there should be no openings that a ball with a four-inch diameter can pass through. So, when you find railings with large gaps you just write it up as a potential safety hazard. You don't have to say it's a code violation. You say something like what I said in this situation: "The railings throughout the house have openings that are large enough for a small child to fit through or become entangled in. Properly modifying or replacing these railings is recommended to prevent an injury or fatality."

Ruling: Fact. I don't ever use code to reference my reason for writing something up in a report. The only time I might mention the word is when advising someone to repair something to the current code. For example, "No functioning smoke detector was found. Updating the house to the current code is recommended for safety." If you have a hard time avoiding the word "code," just use one of the numerous substitutes—phrases like "modern-day standards," "best building practices," "traditional building methodology," "standardly recognized building practices," etc.

Many home inspectors also recommend not using fancy tools. When you pull out a moisture meter to detect water damage in floors, or an infrared camera to discover problems in a wall, you change the buyer's idea about what you are able to find. These things elevate the expectations of your buyers to a dangerous level. Your contract should refer to standards of practice and repeatedly hammer home the fact that you are performing a visual inspection of the readily accessible portions of the house. Breaking out a bunch of tools to see through walls and floors goes directly against this.

Ruling: Fact. I don't carry fancy tools. I know I'm in the minority on this, but in nineteen years of performing my own inspections and running my company I have never had a single callback or complaint regarding an instance where a moisture meter would have helped me. And we're talking about well over 20,000 inspections. So I'm going to keep doing things the way I have been. As for the infrared cameras (and whatever magical technology comes along to help us see through walls in the future), I have a similar feeling. These gizmos just open you up to liability by elevating expectations about what you will find.

Just to be clear—there is a place for these tools. I just don't think that it's during a standard home inspection.

It's very misleading to offer a "free IR scan with your home inspection"—you can't possibly perform a quality IR scan on an entire house in a few hours, let alone perform a home inspection during that same time. Oh, you're only going to focus on suspect areas? Good luck when something comes up in a non-suspect area. Your contract is going to have more holes in it than a road sign in Texas.

Thankfully, this industry has evolved over the years and we've learned a lot about how to protect ourselves from liability. We have great contracts, regular training, and advisement from specialist attorneys, along with standards of practice. All of these things help keep us out of legal trouble. But some home inspectors argue that you should never report on things outside your contracts and standards of practice, or else you open yourself to liability.

Ruling: Folklore. As long as you remember a few key things as you write your report, you can dabble in areas beyond your contract and standards of practice, and still protect yourself.

So, when is it okay to go beyond your contract and standards of practice? When you remind the client in your write-up that there are limitations to your findings. For example, if you see mold in the attic, you can write, "There is a discoloration on some wood surfaces throughout portions of the attic. This is likely a mold or mildew of some type and is the result of an excessive amount of moisture and/or a lack of adequate ventilation.

Identifying specific mold types and determining the extent of any mold problem is beyond the scope of this inspection. Further evaluation from a mold remediation contractor is recommended."

Similarly, if you notice failed thermal pane window seals, you can write, "The large front-facing living room window has a failed thermal pane seal. This condition is evidenced by a haze that cannot be cleaned away from either side. Replacing the glass panel is typically necessary to correct this condition. Depending on the time of year, how recently a seal failure occurred, and the outside weather conditions, a failed seal can be more or less noticeable. It is possible that additional windows are damaged and were not visible at the time of the inspection."

Mold is outside most home inspection standards of practice, since it is an environmental concern. Failed window seals are disclaimed in my contract, because they can't always be seen. While in both cases I went outside either my contract or the standards of practice, I also reminded the client of that, or otherwise protected myself. As a result, clients would be hard-pressed to ever mount much of a case against me.

The other relevant consideration here: reasonableness. When inspecting the attic, you well may be the first person to visit the area in the last twenty years. Any reasonable inspector understands that you must say something about the black mold all over the underside of the roof deck. I have personally talked to inspectors who bury their head in the sand and try to fight this, saying that mold is outside of the standards of practice—but wow, what a long and difficult road that can turn into. The failed window seals just get talked about so much with buyers and agents, and having chalky cloudy window is pretty undesirable. Again, I only have it in my contract because you can't always see them. Another way of looking at it is that I want the absence of any information in the report to fall in my favor rather than against me. The biggest takeaway from this is that to be a successful home inspector you must always remember to protect yourself with the write-up, especially when you start to wander outside the friendly confines of your contract and standards of practice.

PART THREE

PERFORMING AN INSPECTION

CHAPTER FIFTEEN

PERSONALITY, NOT TECHNICAL KNOWLEDGE

IF YOU'RE SCARED OF YOUR COMPETITION YOU need to look in the mirror, because there is a good chance you have more to worry about than that. I have respect for some of my competition, but I'm not scared of them. Rather than let them intimidate me, I do everything I can to learn about them. I listen when buyers or agents talk about other inspection experiences. I look at trucks on the freeway with inspection logos to get an idea of what services others are offering and to learn of competition to go research. And I get my hands on as many competitors' reports as often as I can. If you expect to compete in your area and get the call ahead of the next guy, you must listen to the people making that call, and give them what they want and need.

If I could offer a single piece of advice to someone considering a career as a home inspector, it would be this: you must have a personality that sells! "Sells what?" you ask. In a word, you are selling *yourself* as a home inspector who is trustworthy, hardworking, helpful, dedicated, etc. I could go on with these resume clichés for pages, but I'm sure you get the point. Just knowing a lot about houses isn't enough.

You could be the most knowledgeable person on the planet with every technical aspect of home inspections swirling around in your head and, unfortunately for the public doing the hiring, you will fail miserably and won't be in business if you don't have an attitude and personality that engages people.

Here's a story of a couple inspectors who have worked with and for me over the years—let's call them Ronald and Cliff, though the names have been changed to protect the innocent.

Ronald was one of the most memorable, nicest people I have ever met. He had a smile that could warm a room in seconds, and a personality to match. Ronald came to home inspections fairly late in life but he quickly caught on and worked as much as he wanted up until his retirement. In fact, more than ten years after he retired I still got phone calls from past clients looking for him, because their experience was so positive.

Now for the skinny on Ronald's technical skills: He'd be the first to tell you he really had no idea what he was doing performing home inspections. He'd sit at meetings with a glazed look in his eyes as some technical topic was going around, and would joke about his lack of knowledge on the subject later. In part, that was Ronald's personality—he was very self-deprecating. But he also wasn't a whiz when it came to the nuts and bolts of inspections. On more than one occasion I found myself in a driveway, checkbook in hand, paying to fix something Ronald should have reported on at the inspection. Still, Ronald more than made up for his technical shortcomings by leaving a lasting impression on people, and selling himself as a caring and genuine person. From this, most people jumped to the conclusion that Ronald was also a great inspector.

Now for Cliff. On the inside, Cliff was as great of a person as Ronald. I really enjoyed getting to know him, and working with him for years. I will still call him from time to time with a technical question, or just to catch up. Cliff had worked as a remodeler, torn houses down and rebuilt them, and is without a doubt the best hands-on, all-around contractor I have ever known. He could install trim throughout a room in what seemed like minutes with precision accuracy. He could hang doors, install siding, wire rooms, plumb fixtures, repair rotted decks—at times I was convinced he was equal parts magician and contractor. Cliff was a true master of all aspects of construction, and his knowledge was absolutely amazing.

Now for the skinny on Cliff's people skills: He did okay as an inspector, but ultimately didn't catch on, because he just wasn't outgoing enough, and didn't sell himself. Cliff ended up going back to being a remodeler, and now he does very well. Being an inspector just wasn't for him, and it's really too bad. If the world worked the way it should, a guy like Cliff would be the most successful inspector. But people are often sold on delivery rather than substance. This is especially true with a home inspection, where there is so much emotion involved. People are making a huge decision, and they need help. They need someone to make them feel good (or not) about a house.

I regularly see very skilled inspectors complaining about a lack of business, and how that's a sign of injustice in the world. This group usually resorts to accusing their competition of writing "softball" reports or pandering to real estate agents to get referrals. But to reiterate my initial point—if you are scared of your competition, you might need to look in the mirror.

In the end, a truly successful inspector will have a good balance of technical knowledge and people skills. If you have too much missing from either side you likely won't ever become a successful home inspector.

CHAPTER SIXTEEN

NINE-TO-FIVE
HOME INSPECTIONS

WHEN I TALK ABOUT BEING A SUCCESSFUL HOME inspector, the first thing I'll admit is that there is a lot more to life than doing home inspections. To me, being successful involves having balance in my life, and other things to do outside of work. The ability to set my own schedule and essentially work whenever I want to is one of the more appealing parts of this career.

Of course, work is important. But always remember it is just a vehicle to allow you to spend time with things that are really important—like a wife and kids, for example. I don't enjoy sitting in front of my computer at 10 p.m., trying to get a report out. I never considered myself successful during the years I spent doing things like that.

To be a successful home inspector, you must find a way to get your reports done and put the finishing touches on your day at a reasonable hour. Nowadays, there are dozens of report-writing software systems, and even more electronic gadgets and gizmos to assist you. Do a lot of research and decide what works best for you. Of course, no gizmo, gadget, or bank of report terms will actually write the report for you. There is a certain amount of time that you must spend customizing a report for each house, in order to turn out a quality product.

My system has changed around over the years, but I generally just set up a laptop on the kitchen counter and stop by several times throughout the inspection to input data. I've found that standing in front of the computer for long periods of time is not perceived too well by clients, so I try to stand at it for no more than fifteen minutes or so at a time. For most houses, a couple stops like that at the computer and I can leave the house with the report nearly done. In most cases, when I walk out of a house all I have to do is proofread and maybe play around a bit with the pictures. Then the report is out the door to my client, and I'm in the door to hang out with my family or do other non-work stuff.

Another benefit of working on the report onsite is that you can fill some time, so you are there long enough to please the buyer. We all move at different speeds, and there is a wide variety in how long it takes an inspector to get through an inspection. Having come from the restaurant business, where speed and efficiency are key, I can get through some houses very quickly. Having my laptop set up and being able to work on the report really helps fill some time while onsite and ultimately helps me get the report out more quickly.

In the end, to become truly successful and make home inspections a long and enjoyable career, you must find a way to be done at the end of a normal work day.

CHAPTER SEVENTEEN

GET A ROUTINE

IN ORDER TO CONSISTENTLY COMPLETE QUALITY home inspections and give your buyer what they are paying you for, you must develop a solid routine. That way you will get through a house in an efficient manner.

Getting an inspection done in a reasonable amount of time isn't just so you can get on to other inspections, or other things in your life. It's for your clients, the sellers, and the real estate agents as well. There is no quicker way to get a bad reputation than being known as "the six-hour inspector." There is a fine line between being thorough and being incompetent. Thorough is a few hours onsite, getting a good look at a house, talking a lot with your buyer, and either handing off a report when you're done or getting one to them shortly afterwards. Incompetent is five or more hours onsite, wandering around thinking if you are there long enough it will convince everyone you know what you are doing. One of the more common complaints I've received about newer employees over the years is that they took too long or seemed lost. You must have a sense of purpose as you go through the house, and always be on your way somewhere. It's obvious when you're wandering and wondering where to go next, and people will lose faith in you in an instant.

Every inspection is different, but an average house should take you in the neighborhood of two and a half hours. Of course, there are huge old houses in terrible condition that take longer than that, and small condo interior inspections that can be done in less than an hour. The two and a half hour thing is a rough guideline. In order to get through an inspection in an acceptable amount of time, and to be sure you cover all the necessary items, you must have a pattern that you follow on every inspection.

There are many different ways to physically move through an inspection, and I'm not going to tell you how it must be done. Some of the newer software programs even have ways for you to collect data and write portions of the report as you're walking through the house. Many "old timer" inspectors I know still use good old-fashioned pencil and paper, or talk into a tape recorder. My best advice is to play around with your routine and just gravitate toward what works best for you. Just because you paid a huge amount of money for some reporting software, don't think that you have to use it forever. Just remember to inspect all of the components of the house, and don't lose any of your data between the point you collect it and when you write your report.

When you first start out, it is important to keep a strict routine just to be sure you get to everything—so I'd highly recommend a written list, or something similar. As you get more experienced, you will start to have a running checklist in your head, and will just know when you've seen everything. At least that's the way it has worked for me over the years. Also, when you get more experienced, it becomes easier to get pulled away from your routine and still finish the inspection without forgetting things. Buyers love to pull you away to show you things they are concerned about, so prepare for that. Also, even though sellers are asked to leave the house during the inspection, you will often encounter sleeping teenagers in certain rooms, dogs being shuffled around, and a whole bunch of other things that keep you from going through a house as you'd like.

Here's my routine. I typically arrive between fifteen and thirty minutes early. The first thing is to determine whether the house is vacant or there is someone home. Once that's out of the way and I have access to the outside of the property, I complete most of the exterior and roof inspection. Starting on the outside also has the benefit of enabling me to get some work done while waiting for the ever-late real estate agent to open up the house.

Once inside, I set up my laptop on the kitchen counter or some other convenient place and then get to inspecting. The first step in my interior inspection is to do a preliminary walk-through of the house to identify attic and underfloor crawlspace access points, and just get a general sense of how I'm going to get through the house. I'll usually turn on the furnace, air-conditioning system, or blower fan (depending on the time of year), to check airflow to each room.

After the walk-through I do what I call "the mechanicals." I check the electric panel, the furnace, and the water heater. Next, I get into the attic(s). This is roughly the halfway point of my inspection, and I'll often take a break and start inputting things into the laptop if the conditions permit. This is also a good time to give your client an update of what you are finding.

Each time I stop at the computer, I really aim to get "caught up." This means I write up every problem I've come across so far. After talking into a tape recorder for years and then taking hand written notes for a short period, I have settled into relying on my camera for documenting problems I need to address in the report. I take a picture of everything I need to write about. In many cases, the picture is never intended to make the report, but is just a way to jog my memory when I'm completing the report. The last thing I do before proofreading a report is to run through all my photos a couple times, to be sure I have everything in the report that I need.

Okay, now that I'm caught up with the first part of the inspection, I set out on the interior of the house for my detailed room-by-room inspection. By now most of the large-scale items have been addressed. Things that come up on the inside of the house are generally more minor but can still be significant. I start in the kitchen and go through the house systematically, one room at a time. *Always* walk back under plumbing fixtures on upper levels to check for leaks. Main floor plumbing fixtures get checked from the crawlspace. Yes, here in Oregon most houses have underfloor crawlspaces. You lucky guys in most of California, Arizona, Texas, Florida and many other areas probably think I have the worst deal in the world, having to crawl under every house. I'd counter by reminding you we don't have rattlesnakes and scorpions—so it's hard to say who the winner is.

The last thing I do is crawl under the house. Honestly, the main thing I dislike about doing the crawl space is putting on coveralls, gloves, and a mask, and then fighting my way through the closet of dirty shoes to get to the hatch. Once you're in, most crawlspaces really aren't too bad. Of course, there are exceptions.

After exiting the crawlspace and shaking the spider webs out of my hair I'll pack up most of my things and try to catch a couple minutes at the computer. Of course, I also must wrap up with my client and give them the "walk-through" that many agents request. Often, the client sticks pretty close throughout the inspection, so there isn't much need for a walk-through, but a summary can be helpful. The end of the inspection is also a great time for a little chitchat with agents and buyers. I'll often use that time to learn a bit more about the buyer. I'll ask things like where they are moving from, and what they do for work. I'll often share some things about myself if they seem interested. Agents I know well also often ask about my family and kids. I have found that spending just a few minutes of time wrapping up the inspection is a great way to leave your client with a good feeling about the process. Of course, I always end by asking my client if they have any questions, and I always tell them they can call or email me later if things come up.

That's pretty much it for physically performing an inspection. Again, your routine will vary. There is no right or wrong way to go about performing a home inspection. Just have a method that ensures you get to everything you need to get to, and look like you know what you are doing as you go.

Now that you're done at the house, how do you put the report together? Well, ideally before you leave the house you already have a good head start on the report. I try to get most or all of the problems written up before leaving, but that's not always possible. In any case, writing up the problems is the first step. Next, be sure you have all of the required physical characteristics about the house documented, either in your notes or in the report itself. Standards of practice require you to document the materials in the house, as well to indicate where you viewed things from. (Like with the roof, for instance. Did you walk on it? Did you look at it from the top of a ladder? With binoculars?) Honestly, I think few people actually read or digest much of the checklist-format stuff in the reports. The average person doesn't care if they have 150- or 200-amp electric service. Many probably don't even know what that means.

With the text and other information in the report, I then put in all the relevant photos and label them. Next, I go through my "notes" (i.e., all the photos I took) to be sure I have everything in the report. Then I give the report a final proofread to be sure I don't look like some fool who can't spell. After that, it's off my plate, and I will hopefully never have to think about this inspection again.

I remember once while sitting in a class, the person presenting was trying to get a feeling for what we do. He asked, "So, you guys only see your clients once, right?" Someone immediately shouted, "Hopefully!" I've always laughed about that because it is so true. The best inspection is one you never have to revisit or think about again.

CHAPTER EIGHTEEN

THE REPORT—
SHORT, SWEET,
AND TO THE POINT

THE HOME INSPECTION REPORT: I COULD EASILY write an entire book about this topic. There are short reports, long reports, handwritten reports, slick and glossy reports with hundreds of photos, and reports written on scraps of paper (really, I've seen this). I'd encourage you to get your hands on every competitor's report you can, and listen to what clients in your area have to say about them. What are their complaints? What do they like? What don't they like? What is attractive here in Oregon may not be appealing in your state, so always keep an ear to the ground. Don't be stubborn and just do it your way. You must be open to tweaking and changing things periodically. For example, do you deliver your reports onsite or do you email them later? You can check into an internet chat room and see inspectors fight about this for days.

Like most things in this business, do what works best for you and your clients. I email my reports after the inspection. Almost always, it's the same day. I know I have some personality quirks—one of them is not being able to rest while someone is waiting on me. The weight feels particularly heavy if the person waiting is about to make a purchase involving hundreds of thousands of dollars. For that reason I get my reports out within a couple hours of the inspection. Realistically, if my morning inspection turns out to have some "challenges," or if my schedule is tight, I may end up sending my morning report out after my afternoon inspection—but I don't like it that way.

As for the report itself, what I've settled into after years of doing this is generally keeping things short and sweet. I don't fill my reports with paragraphs about things I didn't inspect, or things that are outside the scope of a home inspection. These things belong in contracts, not a report. Many inspectors would cringe at this, but I don't have a single pre-written disclaimer in my report. I also don't have a bunch of boilerplate stuff about houses in general. My report is a list of problems specific to a particular house. Clients can find general information about houses anywhere, if that's what they want. The inspection report is only for items that pertain to the house you are inspecting. If you just can't help but pass along general information and disclaimers, put it in an appendix or separate section.

I often read a report from a competitor and struggle to understand how they stay in business. After all, I'm an experienced inspector. You'd think I could pick up an inspection report and be able to determine the condition of a house reasonably quickly, right? Wrong! Most reports have page after page of disclaimers, general information, and lists of things that weren't inspected. Do you really think this is helpful to someone anxiously wanting to learn about their new house?

What other purpose might this stuff serve? Preventing lawsuits? I suppose an argument can be made for this, but I don't buy it. All the report fine print in the world isn't going to stop someone from coming after you if they think you missed something. So is it about creating value perception? Maybe. If a report is sixty pages long and has a few hundred pictures, it must be better than the report that is twenty pages with forty pictures, right? *Wrong!* Homebuyers and real estate agents don't have the time to wade through sixty pages of your ramblings. Shame on you and your ego for thinking they should. Get to the point! Let them know what exactly is wrong with *this* house.

I don't do a summary in my report and haven't been asked for one in the last ten years. My report *is* the summary. Most problems get written up in a sentence or two. The more you write, the more cloudy your actual findings and recommendations become, and the more you open yourself up to liability. There's a common misconception within this industry that if you just keep writing you'll eventually hit upon something that gets you out from under any liability. Realistically, the more you write, the more you sound like an expert on a given problem, and the more you convince your buyers that they don't need to get an expert to sort it out. This is the worst possible outcome. As far as my buyers are concerned, I don't know a lot about anything. I'm at a house to point at problems, not to fully explain them or give exact directions on how to fix them. Leave the Superman cape in the truck and admit there are some things you don't have the super-power to solve. Leave it up to the specialist.

In the end, the report must stand the test of time, so it must be written to protect you. It must also be written in a way that is useful to your client. Many inspectors don't think doing both is possible, but I'm living proof that it is. To be a successful home inspector, always remember these two critical points: Protect yourself in the way you write your report, and inform your client in the way you write your report.

A good write-up of a problem has four points:

- What is it?
- Where is it?
- Why is it a problem?
- What should the buyer do about it?

These four points are required by most standards of practice, whether they are for a given state license or a trade association.

Here are a couple of examples.

An easy one: "The duct from the master bathroom fan has fallen in the attic and should be properly re-attached at its intended roof vent to prevent excessive moisture and related problems."

What is it? A fallen fan duct.

Where is it? In the attic above the master bathroom.

Why is it a problem? If not fixed, it will cause excessive moisture and other problems.

What should the buyer do about it? Re-attach it at its intended roof vent.

This write-up was accomplished with one sentence, but this was a pretty simple one. Most write-ups about problems of any magnitude require at least a couple sentences to cover the points.

Here's one from a recent report of mine—this time there was a little more on the line: "There is a large horizontal crack and indications of lateral outward movement throughout the foundation along the west side of the garage. This is considered to be a failure of the foundation and at some point additional movement and a total failure and collapse could occur. Further evaluation and repair from a qualified engineer or contractor is recommended."

The same four questions are answered: what, where, why it's a problem, and what to do about it. Even though it's a large-ticket item there is no need to write paragraphs about it. It is a failed foundation—end of story, as far as I'm concerned. Get it looked at and fixed by a qualified contractor or engineer. Although the fan duct can be re-hung for $50 and a failed foundation may cost $50,000, the write-up is not much different. Of course, when responding to the inevitable question—"So, how does the house look?"—I'm going to bring up the foundation long before the fan duct. And I'm going to have a bunch of scary pictures of the foundation in the report, where I may only have one (or even none) of the fan duct.

In other words, get to the point quickly. There's no need to write paragraphs about the type of crack, the size, what it might mean, what might have caused it, what could happen in the future, etc. Going into detail on any one of those points can only paint you into a corner and rob your client of the best recommendation. Get a specialist to look at it. It is off your plate and your client is on the road to the best solution.

As I mentioned, I could write a book on home inspection reports so I'm going to leave the subject with what I've covered thus far. Just remember to always keep an open mind and listen to what the people who are paying you are asking for.

CHAPTER NINETEEN

UNDERSTANDING WHAT YOU ARE NOT INSPECTING

DON'T PLUG YOUR EARS AND SCREAM "NAH, NAH, nah" every time you see a swimming pool. It's a hole in the ground with water, and a heater, and a filter system. It's not complicated. You certainly aren't going to inspect it, but there's no law saying you can't at least have a basic understanding of how it works and be able to talk to your client about it. Just because you acknowledge the presence of something you aren't inspecting doesn't mean you're going to be sued out of business.

In this profession, don't be scared, be aware. Being scared changes what you do and, in the end, will cost you. Be constantly aware of what's around you and embrace it. Talk to your clients about the pool. "Oh, that's great— you're getting a place with a pool. Your kids are a great age to really enjoy it. I wish I had a pool growing up. Now, keep in mind my inspection doesn't cover any aspect of swimming pools. My best advice would be to ask the sellers who they have maintain it, and give them a call to see what they recommend. Also, I know a really good pool inspection company. Let me know if you'd like their name and number."

Then, depending on how you feel about the client and how much they really listened, you can put a note in the report that "No aspect of the swimming pool or any related equipment was evaluated as part of this inspection." It's really just that easy.

The alternative, which I don't recommend, is to say something like this: "What's that? This house has a pool? I don't inspect them!" This is not very helpful. Too many inspectors take this approach with the pool, the hot tub, the sprinkler system, the low-voltage lights, asbestos, radon, lead paint, and everything else that you *don't* inspect according to the relevant standards of practice. Buyers are left wondering what actually *did* get inspected. More problematically, they're left feeling like you don't really care. To be a successful home inspector you are equal parts inspector, consultant, and therapist. You absolutely must leave people with the impression that you care about them and the purchase they are going to be paying off for the next thirty years.

There's no harm in talking to people about things you aren't inspecting. Listen to your clients—then guide them, and let them know what they should do. In most cases I find it has nothing to do with money. Most people would gladly pay more to have other things inspected. They just need to know what you are covering and what else they should look into.

To be a successful home inspector, use the list of things outside the scope of your inspection to your advantage.

Coaching clients through this part of their purchase will leave a lasting impression that you care, and the few minutes you spend will come back to you tenfold.

CHAPTER TWENTY

PEOPLE ARE ONLY PAYING FOR SO MUCH OF YOUR TIME

SOME HOUSES ARE A MESS. AND THAT'S BEING nice.

Realistically, a *lot* of houses are a mess. Any Saturday morning, if you want a good laugh, stroll down to your local Home Depot or Lowe's and look around a bit. Not at the products on the shelves, but at the people scratching their heads in the aisles. I'm comfortable going out on a limb stating that most of the shoppers at the local big box store are in over their heads on a given project. The rest—the ones who know what they are doing—are the professionals, and they usually shop Monday through Friday. On a given Saturday morning, the percentage of people properly performing a technical project on their house is in the single digits. The low single digits.

In walking through your local big box store you'll find in the checkout line guys with drywall screws, fence rail brackets, framing lumber, and corrugated fiberglass shed roof panels. But unless they're working on four different and totally unrelated projects, they're headed home to construct a crappy lean-to patio cover (a homeowner favorite, by the way). Yes, homeowner do-it-yourselfers are the lifeblood of a home inspection company. Without them—if only experienced, licensed contractors worked on houses—most of us probably wouldn't have jobs.

So, now that we know what homeowners do to their houses, what can we as inspectors do about it? I'm going to give you a phrase that is worth more than a hundred times what you paid for this book: "Further evaluation of all aspects of the [fill in your item or system here] by a qualified contractor is recommended." This quick little sentence is quite powerful—I don't just throw it out on every inspection. "All aspects" gets you out from under any responsibility of a problem with that system. Unfortunately, this is necessary fairly frequently. You will find houses that have past work and repairs so beyond anything you can sort out it will make your head spin. I honestly don't know how the list-every-little-problem Superman inspector does it. Maybe I've just been doing this too long and see too much. I regularly find houses that literally have more specific problems than I could write up in six hours. And I type really fast.

To be a successful home inspector, you must value your time and realize that people are only paying you for so much of it. I am fine pointing out specific problems up to a pretty detailed level, but there are limits. You must learn to accept that often there are conditions beyond what you can specifically sort out in a few hours. Also, there is added liability in trying to list every single problem. I disagree with a lot of really accomplished inspectors on this one, but I'm not budging. When you list every single problem in a house, all it takes is a contractor to find one more, and your phone will ring with someone complaining. For what it's worth, the newly discovered problem doesn't even have to exist. Some contractor just has to say it does. Yep—some of the worst phone calls and most time-consuming flurries of chaos I've been part of have occurred when a contractor disagrees with some finding, or has more to add, or sees things differently, or just hates inspectors and wants to be a pain in the ass. "Further evaluation of all aspects" saves me time onsite, reduces my liability, makes my phone ring less and, most importantly, gets my clients the best information in a concise manner.

One final plea to Superman: As you are at your computer at 9 p.m., putting picture number seventy-three into a sub-sub-section on electrical wiring deep in your report, I just put my kids to bed after cooking dinner and spending a couple hours doing homework with them, and having quality family time. My two reports for the day were sent off hours ago. Take off the Superman cape and stop trying to solve all the world's problems. You will never be able to discover and list all of the problems in the really bad houses. Trying to do so not only elevates your liability but is a total waste of your precious time. Type three or four sentences with some solid recommendations, give them up to ten pictures of problems, and be done with it. Priding yourself on two-hundred picture reports is just crazy. Are you a photographer or a home inspector? Nobody reads that far into the report anyway.

After a point early on you are just feeding your own ego and trying to convince people you are great.

To be a successful home inspector, you must be able to convey useful information in an efficient package. Get to the point and save yourself time and effort. Most of all, use the time you save to do something for yourself. What's the point in killing yourself crawling under a house through dead cats if you don't have time for more important things at the end of the day?

CHAPTER TWENTY-ONE

DATING HOUSES AND THE SYSTEMS WITHIN THEM

AS A HOUSE SITS SILENTLY, IT IS ACTUALLY screaming information at you if you listen closely. Of course, you need to listen with your eyes. If you do, there are hundreds if not thousands of little tips a house gives about its age and the age when certain work was performed.

For example, let's say you walk into a house that was reported to have been built in 1987 and when you go into the basement you find galvanized steel water piping and a fuse box. The last fuse panels were installed somewhere around the 1950s and the last of the galvanized supply piping was typically put into service in the mid- to late-70s, so the 1987 date can't be correct. In this case, the most likely scenario is that the house was either marketed by the listing agent with an incorrect date, or the incorrect date was called in when the inspection appointment was made (both very common occurrences). Less often, houses are listed at the county assessor's office with an incorrect build date. What I have found is that when the house is incorrectly listed with the assessor, it was often relocated to its current position at the date they have recorded (years ago, when freeways were built, it was common to move houses that were in the way).

It is important to communicate with your client when a house has a different age than what was reported. I suppose technically as inspectors we aren't responsible for this, but you can avoid a headache down the road by letting a buyer know that he or she may have been given some incorrect information. In my opinion, houses are always being made better. Energy efficiency, safety, convenience, earthquake stability, and a whole host of other things improve as time moves on. Many homebuyers are aware of this, and the age of the house factors into their decision.

Aside from the actual build date it is good to be aware of what materials and practices were common at what times to help you gauge when a certain repair or replacement was performed. This can also help determine when an addition might have been put on a house.

So, without further ado, here are some things I use to "date stamp" the house I am looking at:

- Thermal pane windows can be found as original equipment as early as the 1950s in nicer homes, but are most often found starting in the mid/late 70s.

- Wood windows gave way to metal in the 50s.

- Metal windows gave way to vinyl in the early 90s.

- Thermal pane windows often have a date stamp

on the metal jamb between the panes of glass—
this is a great tool to age an addition.

- Toilet tank lids often have date stamps inside.

- Some plumbing vent roof flashings have a date
  stamp, as do some plastic roof vents—these are
  great gauges as to when a roof was installed.

- OSB sheathing has date stamps near the lumber
  grade marks.

- Heating ductwork changed from onsite fabricated
  metal to plastic flex ducting around 1990—date
  stamps can often be found on the newer flex ducting.

- Furnaces, water heaters, air conditioners, refrigera-
  tors, ovens, microwaves, and dishwashers have ID
  tags that will usually have a date—either visible or
  somehow coded within the serial number.

- Fiber cement siding became common around 2000;
  butt joints up until about 2007 were caulked;
  post-2007, the joints have metal or building paper
  flashing strips behind them, so they do not need to
  be sealed (and don't let a painter tell you otherwise).

- Synthetic stucco (EIFS) was common for about ten
  years starting in the mid-90s (and it's a problem,
  to say the least).

- Wood can be found as siding at any time, but as
  it became more expensive over the years you will

see it less often. Various composites or materials such as vinyl are used almost exclusively on new houses today. Also, newer wood tends to have a lot more imperfections (knot holes, etc.) and is easy to identify once you are familiar with it.

- Plywood started being used in residential construction in the 50s.

- True two-inch thick lumber was phased out around 1920. (A modern "two-by-four" is actually a "one-and-a-half-by-three-and-a-half"—I really hope you already knew that).

- Tongue and groove floor decking boards were often run perpendicular to floor joists until sometime in the 20s. Then they were run at a forty-five degree angle across the joists until roughly 1960, when they were turned perpendicular again (I never have understood why they changed this several times).

- GFI outlets were phased into different locations over a long period of time (there are some good charts out there that explain when and where, if you do a little googling).

- Knob and tube wiring was phased out around 1920; the early NM cable (kind of has a snakeskin pattern) was phased out around 1970 in favor of plastic-sheathed cable; romex cable started being color-coded around 2000 (what a great idea—why had no one thought of this previously?); newer electric cables have dates stamped on them. This is particularly convenient when a seller states that the previous owner put the addition on the house but all the wiring is dated three years after he bought the house.

- Aluminum branch circuit wiring can be found anywhere from the late 60s through the late 70s.

- Galvanized steel-supply piping was used until the mid- to late-70s; copper from that time until the late 90s; CPVC plastic for a few years. Now PEX is pretty much all you see—date stamps can often be found on the plastic materials.

- Cast-iron waste lines and galvanized drain lines (often with cast-iron fittings) were used up until the mid-70s, and ABS black plastic or white PVC plastic have been used ever since—often with date stamps.

- Crawlspace foundation vent screens were set into the concrete up until about 1950; wood frames were installed until the early 70s; plastic vents have been standard ever since then. This is particularly good information when looking at a foundation and suspecting an addition.

- Gas water heaters in garages had to be elevated at least eighteen inches off the garage floor up until the early- to mid-2000s. Today they can sit directly on the floor (by the way, this has to do with an open flame near the garage floor being a spark-ignition hazard—newer gas water heaters are manufactured with closed burners).

- Vapor barriers in crawlspaces were often left out completely prior to 1950. From 1950-1970 they were clear. Ever since then they have been black. So, let's say you go into a 1920s crawlspace with a black vapor barrier. Backtracking in time, there is a really good chance that that crawlspace sat with no vapor barrier for over fifty years, and the wood is eaten by termites and wood-boring beetles. This is a perfect example of how you use the materials you find and the dates they infer to be a better inspector.

- Interior finish materials offer endless clues about when they were installed. I'm not going to go into detail but I would highly recommend you get familiar with basic design changes over the years. This will really help you get a sense for how everything stacks up. For example it's common to go into a 1920s house that was totally gutted and redone in the 1970s. Knowing what materials were used when can really help you anticipate what is in between the two ends of the plumbing or wiring you see. Speaking of redoing things, I once inspected an old house that had had very little done to it over the years. The only major upgrades I could find were the siding and the wiring. As the saying goes—timing is everything. The wiring was updated with aluminum (terrible material for

light and outlet circuit wiring—see the Consumer
Product Safety Commission for more information)
and the siding with LP wood composite (the subject of a class action lawsuit and easily the biggest
blunder we've made in construction over the last
forty years.)

- Metal hangers have changed over the years. I
usually don't find dates on them but you should
take note of what older and newer ones look like.
Gusset plates on trusses have changed over the
years as well. Finding them made from plywood is
common up until roughly 1960. They are mostly
metal after that. Similar to the metal hangers, the
look of metal gussets has changed over the years.

- Concrete has very different characteristics if you
really look close. On newer installs you will often
see the imprints from the plywood forms. On older
houses it was common to see the floor decking
boards used as form boards. You will often see concrete residue on the boards as you look up from the
basement (don't mistake it for mold). When you
are in a really old house and it has no residue on
the underside of the floor decking, there is a good
chance the house was built without a foundation.

A final word on the subject: the above list is based
on my experience, and the houses I've seen in my little
slice of the world here in Oregon. I'm sure houses are
built differently elsewhere, and some of my list might be

inaccurate when applied to other areas. My overall point with this is that you need to educate yourself about how houses in your area have changed over the years. Aside from helping you become a better inspector, just being able to point out and talk about the things you see goes a long way toward building your credibility. Listening to the information a house gives you is a key part to becoming a successful home inspector.

CHAPTER TWENTY-TWO

WHAT NOT TO SAY

YOU WILL HEAR REAL ESTATE AGENTS VENTURE down dangerous paths when talking about things. I've heard the following statements (most of them more than once):

"The radon level is high? Oh, it's just a little high. Remember to leave the windows and doors open when you can to get the air moving through the house. It's not really a problem."

"Mold in the attic? Just spray some bleach on it. It will be fine."

While handing out the lead paint disclosure pamphlet, and laughing: "All this really says is to not chew on the paint and you'll be fine."

While standing outside looking at failing (likely lead) paint on an old house: "Just make sure to wear a mask and scrape the paint off when your neighbors aren't home to report you to the city."

When in the presence of conversations like this you have three choices: leave the area, stand there and say nothing, or give the client the best advice. That best advice, of course, is to seek advice from people who are actually qualified to give it. The main thing is to not agree with or confirm the bad advice of the real estate agent. If they want to risk their butt laughing off real problems, so be it. You just can't put yourself in that line of fire. And realistically, as an inspector, your advice carries a lot more weight than the agent's. Buyers are looking to us for advice. I'd estimate that few homebuyers know that environmental items are outside our scope as inspectors. They're often surprised to learn that many state standards of practice specifically forbid inspectors from dealing with environmental items. Offering up wrong opinions on environmental items is particularly bad.

Another thing to resist is badmouthing contractors and what they may have charged for a given job. We once had an inspector look at a receipt on a table and exclaim, "Wow, somebody sure got taken!" What good could possibly come from that? All it's going to do is to add fuel to a fire. Remember, as inspectors we're there to assist people—we want them to feel better by educating them. Along the same lines, try to avoid trashing the job a contractor (or homeowner) did. It's nearly impossible at times, but snickering at someone else's work can be a bad thing. Often, buyers aren't buying the house they want—they're buying the one they can afford. Too much ridicule can turn negative in a hurry.

This one is a no-brainer, but is worth mentioning: Stay far away from politics, religion, and anything along those lines. Even if you happen to agree with a conversation, there is nothing to gain from jumping in. A great time to do a little research on your competition is at a local inspector association conference. It always amazes me how many of the trucks in the parking lot are plastered with political bumper stickers. Depending on where you are, *any* political view you express is going to turn off roughly half of the people you meet.

Don't *ever* let on if you are sick, injured, or otherwise not at the "top of your game." Buyers will remember this the first time something goes wrong, and chalk it up to you not feeling well or not being physically able to get through the house. Realistically, an inspection is only a few hours of time onsite. Do what you have to do to pull yourself together, and get through it with a positive and upbeat attitude. Your clients don't care if you're fighting with your wife, have a leaking pipe of your own at home, have a really bad cold, or have a really bad hangover. As far as they are concerned, you should be completely "on" for every minute of every inspection you ever perform.

Avoid comments about bad neighborhoods, freeway noise, a big ugly building next door, an ugly RV in the next driveway, overhead powerlines, or anything else that may be less than desirable. Trust me, they have already noticed it. By the time they call for an inspection, they've seen the property and don't need you to remind them of the obvious downsides. On the other side of that coin, *do* point out the positives. I have no problem pointing out nice things. Great backyard, nice quiet street, close to a park, popular part of town with lots of shopping nearby, etc. I don't necessarily go out of my way to talk about this stuff, but it can make for nice chitchat while you're waiting for an agent to open up the house.

Don't talk too much about your own house or your own experiences. A little of this is okay when relevant, but home buyers are really focused on one thing—the house they are buying. Over the years I've taken note of people's reactions and changed my approach a lot. As I start to tell a story I can read people's faces, and after about thirty seconds they totally lose interest and start wondering, "What in the world does this have to do with my house?" Keep stories short. Thirty seconds max, and it *must* be relevant to their prospective house. Going on and on about your past work history, projects you've done, and things of that sort is a common complaint about inspectors. Always remember you are there to look at a house and talk to the buyer. Stay focused. (I'm talking to myself as much as anyone here, as I can get dragged into wayward conversations sometimes!)

Over the years as you build up a client base you will undoubtedly become very familiar and comfortable with some real estate agents and even develop friendships outside of inspections. Of course, when this occurs, all the rules about stories and whatnot are off the table, at certain times. Always be mindful of your buyer and the time at the inspection. It's against everyone's best interest to have a buyer think you are too close with their agent. Always remember this as you are inspecting, especially when the agent doesn't keep the line so well-defined.

CHAPTER TWENTY-THREE

TAKING YOUR EYE OFF THE BALL

A HOME INSPECTOR IS A HUMAN BEING. EVEN the best, most accurate, smartest, most energetic home inspector is a human being. And, unfortunately, being a human being brings some less-than-ideal characteristics to the table. To be a successful home inspector, you must realize when your moments of weakness are the most likely to occur, and do your best to minimize them. Most importantly, you must take steps to limit the damage that comes from having your eyes off the ball.

This book is a compilation of my career as a home inspector, and was started after roughly twelve years of performing inspections on a full-time basis. As I write this, I'm just over sixteen years in, and find myself in the uncomfortable position of being under fire for some past inspections. No matter how long you have been in this business, you will have days, weeks, or even months like this. Errors and misses can be the result of fatigue, physical or health problems, bad luck, timing, things occurring in your personal life, or a combination of these or other things. While I spend a lot of time and effort looking in the rearview mirror (and writing checks to cover mistakes), I know it's more important to get up, dust myself off, and not make the same mistake for the same reason twice.

So, with that prelude, here are my best recommendations to help keep your eye on the ball.

By far my biggest weakness is when I find something big. Let's say I go into a crawlspace and do a preliminary examination and spot something off in a corner. Sure enough, there is a huge termite infestation. I crawl through cat crap and slither under heating ducts, only to find what I suspected—massive destruction. Then it's around to the other sides of the crawlspace, where more damage is found. I'm snapping pictures and blowing apart termite-infested wood with my screwdriver. I feel like the best home inspector in the world. After all, I've just saved my buyers from making a huge financial mistake. But it's so much fun busting open the termite-infested floor joists that I don't notice the large cracks and failed foundation at one corner of the house.

Fast-forward a couple months, after the seller paid $20K to fix the bug damage and my clients move in. Their landscaper is pruning the bushes and uncovers a big crack in the foundation. The next call is, of course, to the predatory concrete-jacking contractor who advertises on the radio eight times a day.

I bet you can guess who the next call is to. Yup—they have a bid for a $17,000 foundation repair in their hands, and they are staring squarely at me.

Okay, full disclosure: that last story was largely fictitious. It incorporates some pieces of events that have almost gotten me, but that thankfully I avoided. Still, the pieces are relevant and instructive, which is why I include them here.

Here's one that isn't made up. I inspected a 1950s daylight ranch house, on a lot that sloped sharply down from front to back. Now, I've found my share of foundation failures over the years, but a particularly common one is on the back or side of a garage on a house like this. There is a tall foundation wall, and a lot of load, with a car and a concrete slab in the garage. In this case I fought my way through some bushes, down to the back corner of the garage, and noticed a very slight bow in the concrete. There were also some fairly minor vertical cracks near the corner, and just enough separation to get my attention. So, up to the garage I went.

Of course, the well-meaning seller had moved all of their things from the house into the garage, and specifically along the side of the wall in question. After moving some things, I discovered that the suspect foundation wall had separated roughly one inch from the slab. This was what an engineer would describe as a failure of the foundation, and the fix runs around $25K. Wow, I'm great! I had a good idea of a certain house's weak point, saw a subtle sign of a problem, did some investigation, and it all paid off. I'm the best home inspector in the world again!

At least at that moment I was. What I missed while dancing around the garage in celebration was that the roof framing was completely messed up. It wasn't the worst thing in the world, but it definitely had some problems I should have seen. Basically, over-spanned collar ties had been used for storing a bunch of stuff, and had begun to splinter at some notches that were cut for ties running perpendicularly. By the buyer's admission, the situation had been made worse after I was there and some contractor had been crawling around on them, but I probably still should have seen the problem.

My business partner largely dealt with this complaint. He ended up giving the buyer $1,000 to fix it. I've given a lot of money to people over the years, but for some reason this one never sat right with me. I saved the buyer $25K with a great find on a foundation, but I got my nose rubbed in some marginal framing. The buyer came out $26K ahead and I lost $600 on the deal ($400 for the inspection minus $1,000 for the framing).

To be a successful home inspector, you must always keep your eye on the ball and stay focused. Things will get chaotic, and big problems in houses will pull you in all directions. Keep your eye on the ball at all times. Doing so is why you make the "big bucks."

CHAPTER TWENTY-FOUR

YOU WILL CALL THINGS INCORRECTLY

IT'S JUST STATISTICALLY IMPOSSIBLE TO ALWAYS make the right call. There is no way a home inspector can always be right. In fact, not even close. On every inspection, you are observing and making a judgement call on thousands, if not tens of thousands of things. The main thing is you must minimize and learn from the times you are wrong.

One of the best ways to avoid making incorrect calls is to have a healthy amount of skepticism about how much you know. For example, about sixteen years ago, when I was still a somewhat new inspector, I pulled up to an inspection on a new house. While walking up the driveway, I noticed in the garage that the gas water heater was sitting directly on the floor. It just totally jumped out to me, since gas water heaters had always been required to be elevated eighteen inches. As I set my stuff down near the front door, I could see inside a few other garages on the new houses, and they were all the same way. When the buyer walked up I was getting ready to run my mouth about how some plumber had totally botched the entire neighborhood of water heaters. Something made me hold back, and I'm glad I did. It turned out they had just started manufacturing water heaters so they were safe to sit directly on the garage floor. I learned that after making a phone call from the house (this was around 2003, and there were no smart phones with internet yet).

Along these lines, I also remember when one of my employees called me to ask for some clarification on a boiler-fed heating system in a unit he was inspecting in a relatively new, luxury high-rise building. Having been in the building several times, I explained the heating system to him pretty quickly. Almost in passing, he laughingly mentioned that the entire place was wired wrong—there were no ground wires in the panel at all. He just couldn't believe it and was having a good chuckle.

Let's stand back and look at the possibilities here. Either an entire high-rise was wired wrong by dozens of commercial electricians and that wrong wiring was missed by all the city inspectors and superintendents ... *or* my inspector didn't understand that commercial buildings often don't contain traditional romex wiring and get their ground from metal conduit. I quickly informed him it was the latter, and hoped like hell he hadn't opened his mouth yet to the client and agent. Luckily, he hadn't, and it all ended well as a good learning experience.

The very important lesson with these examples is to keep an open mind and remember that you don't know everything. Rules and codes change, and have regional differences. You might not know something, or you might have gotten some misinformation along the way. One of my favorite things to tell people when they bring up the code is that just when I'm sure I know something, someone pulls out the 600-page book and shows me why I'm wrong. Leaving the door open to the possibility that you are incorrect is an extremely important part of being a successful home inspector. Maybe it's just my nature, but even when I get questioned on something I am absolutely certain I am right about, I will check before responding to be sure there haven't been any recent changes.

So what do you do when you make an incorrect call and are wrong about something? Like the saying goes, beg for forgiveness. If you make a mistake and call something wrong, don't make it worse by refusing to admit it. In the end, you will come out with the respect of everyone if you are honest and upfront. The good news is that the longer you are a home inspector, the more you learn, and the less often you will be wrong. But, when you *are* wrong, do what you have to do to make it right. Call your client and explain how you arrived at the error. Re-issue the report with updated and correct information. This may involve a return trip to the property, to investigate and learn.

The sort of thing I am talking about here usually happens very soon after your inspection, when some of your findings are being investigated by contractors. When you call things wrong or miss things, and your client buys the house before discovering your mistake, it can open a whole other can of worms—a can that usually has your checkbook sitting somewhere near it.

CHAPTER TWENTY-FIVE

BE NICE TO
THE SELLER
AND THEIR HOUSE

NOTHING IN THIS CHAPTER HAS ANYTHING TO DO with what gets written on the pages of my reports. But it has plenty to do with being a successful home inspector, and an overall good human being.

From the instant you arrive at a house, you are part of the negotiation and purchase process taking place between its buyer and seller. Many times I pull up at an inspection and notice that the seller is hanging out. I go out of my way to meet them—to introduce myself or catch up with where things are, and to ask them if they have any questions about the process.

Selling a house is not easy. There are many reasons people move, and many of them are not positive, so sellers are often on edge to begin with. Of course, my ultimate responsibility is to help my buyer, and selling myself as a reasonable person to the current homeowner is often a huge step toward that end. I usually find sellers to be very receptive and helpful. They'll often show you tricks about getting into attics and crawlspaces, and give you other pieces of information about the house that you otherwise would never know. Often, they divulge more than they probably should about things they've done to the house, and problems that have occurred. Of course, I'm listening closely to every word.

It's best if the sellers aren't at home during the inspection, but there have been many times that I've carried out an entire inspection with the seller sitting in the living room. In the end, it really doesn't bother me that much. I just do my thing, taking notes and pictures, and writing up my report.

The thing you must always remember is that while any inspection is just another day at work, you are also walking into more than just another house. You are in someone's home. It's common to be at a house that the seller has lived in for thirty years or more. That's a really long time. Births, deaths, marriages, graduations, and all kinds of other life-changing events have occurred in or near this space over the years—so keep that in mind as you come stomping in. Again, none of this directly impacts the inspection process, but respecting the sanctity of a home will help remind everyone that you care, and will go a long way in helping your client.

When a seller sees an unfavorable part of your report, it's easy for them to chalk it up to you being unreasonable. If you're cold to them in the driveway, or if you duck around the side of the house as they walk out, they're even more likely to jump to that conclusion. But if they remember you being a decent person, they'll have a much harder time discounting your negative findings. Don't be afraid of the seller. Just because you treat them well doesn't mean you can't be honest and objective, and perform a good inspection on their house for your client.

The other side of respecting the seller and the house is to be clean and, like your grandmother likely told you, "Leave a place better than you found it." That is such an old cliché that for most of us it just flies in one ear and out the other. As you do inspections and are in and out of people's houses, you must always respect the fact that you are in someone's *home*. Again, it's just another day at work for you, but I promise most sellers come home knowing the inspection has taken place, and they go on a hunt—looking for dirt tracked on carpets, appliances left on, doors left open, missing possessions, and anything else they can get upset about. Always remember what every individual seller is going through. This is not just another day for them. This is the day that can make or break them financially.

Here are a few tricks I've learned over the years.

Always carry a towel and slippers. The towel goes down at the front door for everyone to wipe their feet on, and the slippers are for running around inside. Even if you never put the slippers on, just holding them as you walk up to the house shows you care. And the uses for the towel you drop at the front door go far beyond people wiping their feet. Leaking dishwashers, showerheads spraying out of the tub, drying off after testing a tub with the shower diverter engaged (you haven't inspected more than a dozen houses if this hasn't happened to you), and swatting flies are just a few of its uses.

Leave a note after testing the dishwasher. You'll often find a half load of dirty dishes in there, so why not just throw some soap in, since you have to run it anyway? I regularly write, *Ran dishwasher through a normal cycle with soap* on the back of one of my business cards, and leave it on the kitchen counter (a little extra marketing never hurts, of course).

Vacuum and sweep whenever possible. In those nice, clean houses that you can tell were cleaned just prior to the inspection, return the favor and run through the traffic areas with the vacuum if there's one handy. Over the years, I've had several listing agents call me for no other reason than to ask if I had actually vacuumed—and to thank me when they found out I had. Marketing that good usually costs hundreds of dollars—in these cases, it was free, and took me less than ten minutes.

There are plenty of home inspectors who treat sellers as enemies—barely talking to them, and leaving their houses in shambles. Some of these inspectors might even stay busy when the market is hot. But to be a successful home inspector you must put yourself in the seller's shoes, and treat them and their home with respect during all phases of your inspection.

I'll end this chapter with a good story (kind of) about what I thought was a seller's house.

I can't remember exactly, but this must have been at least sixteen years ago, so I would have been in my first few years as an inspector. I was out in a rural area and was trying to find the address. When you're out in the country, all you really have to go on are numbers on mailboxes, so it can be a bit of a struggle. I finally arrived at the property, got my stuff out of the truck, and got ready.

As is often the case, I was about twenty minutes early, so the fact that there was no buyer or agent was no problem.

I rang the doorbell and no one answered, so I started my inspection. I heard a lawnmower running nearby but didn't think much of it. Again, I was out in the country, and the houses were on some really big pieces of property, so hearing someone mowing the lawn wasn't surprising. I set up my ladder at the gutter and hopped up on the roof. As I poked my head above the ridge, I noticed the lawnmower noise stop—the guy who had been cutting the grass was walking briskly up to the back of the house. I'll always remember his exact words: "Can I ask you just what in the hell you are doing on my roof?"

I thought about how the numbers on the mailbox had been pretty hard to read (was that a "seven" or a "one"?). And how there was no *for sale* sign in the yard. And, oh yeah, the house I was looking for was only a few years old, and the roof I was on had clearly been around for much longer than that.

How about that? I was at the wrong house, and on the wrong roof!

The homeowner was actually pretty nice about the situation (after all, he didn't shoot me). I told him his roof looked pretty good but was showing its age. I also told him he might want to get some new numbers for his mailbox. Of course, I apologized profusely, and we both got a laugh out of it. In situations like this I find having a good attitude can really help.

CHAPTER TWENTY-SIX

BREAKING THINGS IN HOUSES

YOU WILL BREAK THINGS IN A HOUSE YOU ARE inspecting. This is different from when something you are inspecting breaks. When a garage door opener blows up and starts smoking after you push the button, it "failed under testing." You are not liable for this. Your job is to test the components of the house. If you just happen to be the poor sucker operating something under normal conditions and it breaks, that's not your problem—though it will usually take numerous phone calls and emails to the homeowner to explain this. Testing things is exactly what you are there to do.

However, catching the flashlight hanging off your belt on a lamp, and then knocking the lamp over and breaking a window—that's your fault, and you are liable for the damage. Remember the distinction between these two types of events—I can promise you people will try to hold you accountable when things fail under testing.

When I think back to the number of houses I've

inspected and the amount of damage that I've caused, it's actually pretty amazing. I personally have paid out less than $1,000 in nineteen years for things I broke. Most of the things have been such a fluke that I still get my money's worth just laughing about it. I know I've been really lucky. I've had some really close calls. Just be aware and be careful—that's about all the advice I can give you. It's one of those things that if you worry too much you'll probably make it worse.

Here are a couple examples.

The first one actually didn't result in me breaking anything, but sure came close—I was in a basement of a nice house with a bunch of wine racks. I spun around and knocked into one of them, and it was like something from a circus. As I tried to catch the bottles, I knocked more off the racks. I stuck out my feet to cushion the fall onto the concrete floor, pinning bottles against the wall with my knee and just doing everything I could to stop the carnage. When the dust settled and everything stopped moving I was absolutely amazed that nothing had broken. An agent and buyer were standing across the basement and just broke out in laughter. Like my grandfather once said, "Sometimes it's better to lucky than good."

This next one was kind of a fluke, but absolutely my fault. I was going through a kitchen trying to test all the outlets. One was covered by a thick wood cutting board, leaning against the wall. I slid it to one side to reach the outlet, and in doing so, I moved it off the wall, leaving it totally on its thin edge. After testing the outlet, I left the cutting board where I had moved it—only a couple of inches from where it had originally been. Fast forward a few minutes, and I was around the peninsula kitchen counter, testing the sliding door. Closing the slider caused the cutting board to fall over. It knocked over a blender, which bounced a glass pitcher off the counter (seemingly in slow motion), making it shatter on the wood floor. Thankfully, there was a good vacuum cleaner nearby, and I went to painstaking measures to ensure I got every glass shard off the floor. I also called the listing agent and admitted what had happened, because the last thing I wanted was to hear about someone with a piece of glass hanging from their foot. Next, I got on Amazon, and overnighted a new blender to the seller before I was even out of the driveway.

Here's one that happened to another guy in my company. He carried a screwdriver in his back pocket, with the blade pointing up. Apparently, he squatted down next to a leather chair, and when standing up he ripped a long gash up the back. When I talk about flukes, this is pretty much what I mean. How could you ever plan for this? It turns out the chair was some old family heirloom, but the seller was pretty reasonable. I think I gave her $300 and she was super happy.

Again, there is just no way to plan for this type of thing. Just assume it's going to happen. You will spend so much time in people's houses that breaking a few things along the way is unavoidable.

CHAPTER TWENTY-SEVEN

GRANDFATHERING

GRANDFATHERING IS ONE OF THE MORE
important concepts you must be aware of to be a successful
home inspector. It's also one of the most often misunder-
stood concepts in construction. Contractors, homeowners,
and real estate agents regularly misunderstand the term
and the concept of grandfathering.

What grandfathering means: A given installation is
perfectly acceptable to remain in place even though the
codes or manufacturer's specs for the installation have
changed. An analogy: airbags are required in new cars,
but that doesn't mean every car without airbags must
be immediately taken off the road. People often want to
know why we don't inspect houses to "the current code."
But in every code cycle there are so many changes that it
would likely be cheaper to tear a house down and start
over than to try to update and change everything to the
new way of doing things. All of the items in an existing

house that have a newer or different installation method are considered to be grandfathered.

What grandfathering *does not* mean: It's an old house that has had a bunch of crappy work done to it over the years, so you can do whatever you want and just say it's grandfathered. New work on an old house must meet the current code. This gets a little murky when you are remodeling and tear into parts of a house or system. People generally must talk to the local building department for clarification as to what can stay and what must be updated. In most cases the folks at the building department are quite logical and helpful. The mere fact that you are taking the time to ask for direction will go a long way.

The concept of grandfathering can get difficult when you start getting contractors involved. A lot of contractors don't know a lot about the way things used to be done. They are trained and tested on the current codebook and are really more installers of equipment than people who are experienced with repairing something built long ago. The problems come about when you call one to ask about an older installation. The first words from their mouth are usually something like, "That doesn't meet current code and must be replaced." To their credit, the first half of that statement is likely true. But the second half is completely false and has been the catalyst for many a phone call tornado session. This is when some buyer, seller, contractor, or agent gets their panties in a wad about something, and everyone madly calls each other, escalating the situation to the boiling point. This occurs regularly, and I often just let my phone explode until the storm has passed. My old boss had a great saying that applies: "A crisis phone call not immediately returned has a way of working itself out."

With most contractors only knowing how to install new equipment—and also having a financial interest in installing this new equipment—I can understand how things fall the way they do. It just gets frustrating when people won't listen to reason.

Along those lines, here's a grandfathering story I suffered through one time. A past client called with a bill for $600 from an electrician. The bill was for installing a new dryer outlet, and he expected me to pay for it. It turned out the kid at Home Depot told him the one in his 1979 house no longer met code. So, he called an electrician—and lo and behold, the electrician agreed, and came out to save him from certain death had he actually plugged in the clothes dryer.

These phone calls are just so funny. Over the years I've become good at all the little tricks. Listen, sympathize, let people vent, understand that most people just want to be heard—blah, blah, blah. In this case, I let this guy go on and on for a while. When it was my turn, he would not hear anything I had to say. In his words, if I didn't at least split the electrician bill with him, he was going to leave bad reviews about my company all over the internet. Getting a little annoyed, I accused him of trying to extort money from me (that was later quoted by him in all of the online reviews, by the way). That really set him off. Looking back, I suppose I should have just paid the guy off, but this whole online review thing just annoys me. Every idiot with a keyboard and an internet connection feels empowered to take out a life's worth of frustration on some poor business owner. Anyway, though I emailed him documentation specifically proving me right, and informing him that clothes dryers are actually shipped with the intention of older three-slot outlets, nothing I could say mattered. He was going down on the ship that he and the money-grubbing electrician and the uneducated kid at Home Depot were all right. He insisted that I owed him money. Honestly, if the guy would have just called me *before* paying the electrician, I would have been happy to just go out and wire the stupid outlet for him. It was an hour job at best, and twenty dollars in parts. I should have been an electrician.

To be a successful home inspector, you must understand the concept of grandfathering, as it will come up often.

CHAPTER TWENTY-EIGHT

DON'T SPEAK TOO SOON OR IT MIGHT COST YOU $400

I CAN'T REMEMBER WHEN I DEVELOPED THIS skill, but I regularly look at a house, and within about fifteen minutes I know most of the large-scale problems I am going to report on. Of course, I do find surprising things when I'm deep into a job, but in most cases I'm not too surprised after my initial walk-through. With this rate of return for time spent, I could probably make a good living just running around doing ten-minute inspections, right? Well, maybe at first—but keep in mind I said *most* of the time I know all the big things. Missing something big even 10% of the time is a pretty bad batting average for an inspector. That would mean once every week or so I'd be leaving out a major defect. I'm sure I'd be hearing back from some clients and their lawyers.

Over the years, I have learned that when I get to a house and do my initial walk-through, I should hold my cards close to my chest. If you haven't guessed by now, one of my problems is talking too much. My wife is more than happy to remind me of this bad habit when it rears its ugly head. The problem as it relates to my inspecting life is that she is never with me. I've had to learn to be aware of my shortcoming, and try my best to keep it under control. I know at times I fail.

Anyway, talking too much early on at the inspection of a really crappy house can cost you. It can cost you because if you reveal too many problems up front, you can talk people right out of the purchase, and make them want to bail on the inspection. I've had many people try to give me a hundred dollars after thirty minutes and say they've changed their mind about the house. They feel that is more than generous. A hundred dollars for half an hour works out to two hundred dollars for an hour, for those of you without a calculator. That's pretty solid money, right?

The problem is I have blocked out time for this inspection, driven to the jobsite, and turned away other work for the same timeslot. I am going to need a lot more than a hundred dollars in compensation. In fact, the $400 that I quoted for the job sounds about right to me. To avoid this problem, I've learned to hold onto the information I gather, and release it slowly—often not until the end the inspection. Just because I have the ability to gather the information about the house quickly doesn't mean the buyer should get a discount. In that instance I'm actually being penalized for being good at my job.

I regularly start an inspection, look around the house a bit and then greet the young first-time buyers in the driveway. I'll pick up on little pieces of information—they got their offer accepted over three others, the seller won't do any repairs, the buyers don't have a dollar to spend over their down payment, Mrs. Buyer is pregnant, and Mr. Buyer is a web programmer who doesn't know the difference between the electric panel and the furnace. I know this transaction is never going anywhere beyond my inspection. Honestly, it pains me to watch the scenario play out. But I go along for a couple reasons. First, I am here to do a job they want done. Their agent is really the one to blame for the circus—an agent should never get people in over their heads like this. Secondly, and most importantly, I want my $400. Yep, I'll say it just like that. It's a cold harsh world out there, and I am one of the nicer people you will meet. But at some point I must look out for myself.

Aside from money, there are other reasons to finish an inspection you start. Sellers will often demand an inspection report in order to let buyers out of a transaction. If you stop early and settle for some meager amount of money you certainly aren't going to go sit at your computer and generate a report. Also, at times, something that is identified as a huge problem is already being repaired. We had this happen one time with a foundation failure. The inspector walked into the house and went right to a large horizontal crack in the foundation—and everyone panicked and wanted to stop the inspection. Luckily, the agent knew that the sellers already had engineers out to look at the problem, and a repair was going to be taking place in the next couple days. As inspectors we just don't always have all the information, and like the saying goes, "the show must go on."

Another time, an inspector was a bit overanxious to get to his vacation, and pretty much talked a lady into stopping the inspection. Fast forward a couple hours and I was on the receiving end of a phone call from an angry husband asking, "What the hell is wrong with you guys? I pay you to inspect a house and you don't finish the job?"

I've given you some pretty good reasons to just put your head down and finish your work. If you still agree to stop an inspection before its completion here is my best advice: Do it for you, not your buyer. (Remember the importance of looking out for yourself and that $400?) If you want to stop the inspection, do it because the house is unsafe, unsanitary, or just such a heap of crap that you honestly can't imagine how you will put it into words. I've had rabid dogs I refused to be around, meth-head tweaker renters, and a house so infested with rats that I was honestly afraid to take a step for fear of squishing one. In these instances you are making a business decision to protect yourself, and that is just fine.

To be a successful home inspector, you must remember you are being paid for your experience and the information you produce. Just because you can produce it quickly at times doesn't mean you should suffer financially.

CHAPTER TWENTY-NINE

SAFETY WHILE
INSPECTING

ONCE DURING A MOMENT OF RESPONSIBILITY and planning I realized I should get some disability insurance. As the owner of my company, I have no coverage from worker's compensation in the event of an injury. I am basically on my own if I fall off a ladder and can't work anymore. When talking to my insurance agent, I was pleased when he told me home inspectors are in a low-risk category for this type of insurance. You can learn a lot by looking at the ratings and costs of different types of insurance. In this case I was pleased to be in a relatively safe occupation.

Still—even though day-to-day inspections are statistically safe, there are plenty of things to watch out for, and of which you must constantly be aware. Walking into a different house or two every day creates an interesting dynamic. You have no familiarity with your surroundings. You are at the mercy of whoever built the house, and whoever has been living there and maintaining it since.

Luckily, as home inspectors, we are generally more aware of surroundings and safety in houses than your average person—so a lot of this stuff comes naturally. In no particular order, here are my recommendations for giving you the best chance of making it home at the end of the day.

LADDERS. This one obviously doesn't need much explanation. You'd better be good and safe with them. Whatever length ladder you decide to use, be sure you can carry it, set it up, and climb on and off in your sleep. In most cases, it's a good idea to specify in your inspection agreement what length of ladder you will be inspecting with, and the limitations that might apply. If you own a multi-inspector company, your worker's comp insurance will likely want to know this as well. There are ladder safety guides and information available just about anywhere. I'll leave it to the ladder companies and their lawyers to give you the best advice—just please be sure you know what you are doing with a ladder, or your career as a home inspector will be short.

ROOFS. Once you know how to use a ladder, most of the time you end up on a roof. Just like ladder length, roof-walking is a hot topic for home inspectors. Some will not get off the ladder, and never set foot on the roof. Some bring hooks, harnesses, and climbing gear. Most inspectors, including myself, walk roofs when it is reasonably safe to do so. We leave the harnesses to the folks climbing Mount Everest. From a ladder, I won't get on a roof with a pitch of 6/12 or higher. But from a window or deck, I'll get on a roof with up to a 8/12 pitch. Rain, snow, moss, or an old roof with loose minerals can keep me limited to the top of the ladder too. The type of roof is also a factor. Here in Oregon, most houses have asphalt composition roofs—that's really what I'm referring to. I rarely ever walk on tile or wood shake roofs. A funny thing that I don't go out of my way to tell my clients is that I'm deathly afraid of heights. Over the years, I've become much more comfortable inspecting and walking roofs. I just don't go too close to the edge, especially when there is a really big drop. Whatever you do, just be sure you always feel comfortable. A key part of every standards of practice I've ever seen is that the inspector does not need to do anything they feel is dangerous. Always keep this in mind, and abide by it.

ATTICS. Don't *ever* put your foot down before verifying you have something to step on. This sounds pretty basic, but I know many inspectors don't follow this rule. You can identify them as the ones who are anxious to tell you the story of the time they fell through a ceiling. Yes, this has happened many times. It could happen to me as well, but I try hard to be careful while walking around above a house. My contract says I won't walk any attic in which the insulation covers the floor framing, but at times I will anyway. The most basic thing to keep you safe is something I call the "sweep and step." Before stepping down, just sweep away the insulation with your foot to be sure what you are going for is, in fact, solid wood framing. If you've ever looked at an attic right before the insulation is installed, you'll realize there are a lot of mechanical lines there. Electric wires, gas lines, and plumbing pipes are just a few things you can encounter. Then there are the loose boards the framers were nice enough to leave for you. Sometimes underfoot, sometimes sitting up in the trusses or rafters. Don't ever trust a piece of framing without first giving it a good tug or step to be sure it's firmly attached.

Okay, time for a story. While I've never fallen out of an attic, I did have a buyer do it once. It was a really tall side attic space and I had walked in a door and about thirty feet down to the far end. There was only a very small amount of insulation, and I just walked along, stepping on every ceiling joist. Before I realized what was going on, I heard a ruckus near the door, and turned around just in time to see my buyer, a very young girl, fall through and catch herself on the framing. She quickly pulled herself up and was very embarrassed about the whole thing. Thankfully, she was unhurt, and actually thought the whole thing was funny. The amazingly lucky part of this situation was that the house was in foreclosure, and had holes in the walls and ceilings everywhere. The agent decided nobody would notice one more, so she didn't say a word about it to anyone. I always write up doors to side attic spaces or crawlspaces, and recommend they be secured closed. This is one of the more glaring shortcomings of the building code, in my opinion. All this focus goes into safety, yet there is no rule against a door that opens into a cavernous pit of a crawlspace, or a side attic space. When my kids were little, it would take them about three minutes to find that and fall in. These spaces often have cute little doors, too, which is a real attraction for a kid.

CRAWLSPACES. I grew up in Portland, Oregon, and have

lived within twenty miles or so of the city my whole life. I have traveled quite a bit, and am well aware of how construction styles vary from region to region—so what I am about to say may not apply to your area. Here in Oregon, 75 percent or more of the houses I inspect have underfloor crawlspaces. I love national conferences and have attended dozens over the years. Without a doubt the most surprising thing to inspectors that I meet from other areas is that nearly every house I inspect has a crawlspace. They think that's the worst plight possible, and often I agree. I wish I could tell a different story, but many crawlspaces are awful.

So, back to the concept that an inspector doesn't have to do anything he or she feels is dangerous. That helps sometimes—but if I didn't ever go in a potentially unsafe crawlspace, I'd never go in any crawlspace. And I'd never work, since word would spread quickly that I was a crawlspace pansy. The reality is that in Oregon you have to suck it up and get into most crawlspaces. There are plenty of hazards—rodent feces, rodent traps, water and electricity, sewage, and probably worst of all, live animals.

Staying safe and healthy in crawlspaces is a topic for an entire chapter (or even an entire book), so I'll try to be efficient here. Look at vent screens as you go around the outside. If any are missing and you see mud tracks in and out, there is a good chance there are animals in there. Proceed slowly, and look around a lot. Trust me, the occupants of the crawlspace see and hear you long before you are aware of them. Just look for those eyes staring back at you.

If you do come face-to-face with an angry raccoon or other animal, remember three things: cover your face, get your legs spun toward it, and start kicking. If you can find a corner to back into, do it. As is true elsewhere in life, prevention is the best cure here, so really look around a lot before venturing too far from the access hatch in a crawlspace that you suspect contains animals. I've been lucky—in seventeen years of inspecting, I've only had a few run-ins, and none were really surprising, because I had always seen missing vent screens or other access points.

With the other hazards in a crawlspace, you just have to use your best judgement. Water, rat crap, and electric wires are all common. One of the best things I ever figured out was to carry a roll of 6-mil black plastic vapor barrier with me. I find that many times the worst part of the crawlspace is at the access hatch. If I can just get in and move around a bit I can actually get through a lot of it while staying dry, or away from the rat crap, or un-electrocuted. Throw down a 10-foot skid of black plastic at the hatch and away you go.

There is another huge benefit to this technique. I regularly see wet or contaminated crawlspaces that I know I'm not going to get very far into. But so much of the home inspection performance involves going under the house. More to the point, if you refuse to go under the house, agents and buyers just go nuts. Even though I already know I'm not going anywhere once I get under there, and I already know what I'm going to write about why I didn't go, there is just a perception thing about actually disappearing into the crawlspace. Throw down the plastic, lie down and take a few pictures, and get out of there. The buyer and agent are both happy because you made an effort. Of course, on the report, be very clear that you didn't get through the crawlspace because of the conditions, and recommend a full evaluation once the place has been cleaned up. By the time the agent and buyer read this, the inspection is over. All they remember is you going under. Perception goes a country mile in being a successful home inspector—trust me.

ELECTRICITY. So, you made it up a ladder, didn't fall off the roof, and weren't mauled to death by a raccoon in the crawlspace. What else could get your wife that life insurance check she dreams of? The answer has to be electricity. An interesting aside here—as inspection claims for missed items go, electrical is almost non-existent. But if you look on message board forums and in continuing education classes, this area is the most popular. Why? It's hard to say, but I suspect it's because electricity is so powerful and potentially dangerous. You must respect it. The number one way people get hurt by electricity is a ground-fault. "What is that?" I'm sure you're asking. A ground fault is when you become the path of electricity to the earth. The most common example of this is standing in a puddle of water and sticking a knife in an outlet.

The best way to think about electricity and safety is that electricity is like a snake always waiting to strike. It's loaded up and ready anytime. You just have to give it a target. Most times you're walking around in rubber-soled shoes on dry surfaces, using heavily insulated outlet testers and equipment with lots of insulation between your wet conductive skin and the charged metal that electrons are waiting to race through. Just don't give that electricity a path. Don't give that snake a target. One of the best tools you can carry is a non-contact voltage sniffer. These are cheap and easy to use, and can sense electricity without even directly contacting it. I carry this pen-like device and use it regularly while crawling through attics and crawl-spaces. Is that old knob and tube wiring abandoned or live? Aside from licking your finger and grabbing it, there is really no way to know unless you have a voltage sniffer.

Electric panels are particularly dangerous, since the largest amount of electricity in the house is concentrated in the panel and nearby. Standard safety protocol dictates safety glasses and gloves when removing panel covers. I should be ashamed to admit this, but I've never used either in removing over 5,000 panel covers. The gloves are in case of a charged cover. Something I learned from day one of inspecting is that you knock on two doors at every inspection. The front door of the house when you arrive (okay, you usually don't do this, but saying it sounds good) and the electric panel door. You hit *every* electric panel you come to with the back of your hand before grabbing the door. This is because if it is charged and you grab it, all of your muscles constrict and you are latched onto that metal door and will very likely not make it home. If you hit it with the back of your hand and it's charged there's a good chance you will make it home—possibly without your eyebrows, but at least alive. The safety glasses are to protect against sparks that fly when the cover is removed. With this in mind I always pull the covers straight toward my face for a couple sec-onds as protection in case of sparks. I realize I'm doing it wrong and unsafe and against the best practices and am putting myself at risk. I very strongly encourage you to follow the standard safety protocol and use all of the best safety equipment and methods available.

PARKING THE TRUCK. I try not to park in driveways unless absolutely necessary. For example, when inspecting a rural property with a driveway that is an eighth of a mile long, it would be somewhat awkward to park on the road. But in most cases I park on a city street with cars screaming by. As my wife will eagerly explain, I have some quirks and hot buttons. One in particular is standing with my back to traffic behind my truck as cars are racing by. I particularly hate standing at the back of my truck while someone is parking or leaving from right behind me, for fear they drive into me. Just do me a favor and pay attention while you are loading or unloading inspection gear at the back of your vehicle. I've read a few stories of people being crushed between vehicles, and it just makes me cringe. I realize I probably have a better chance of being mauled by a bear, struck by lightning, and winning the lottery all on the same day, but it's just one of those things.

Without a doubt there is nothing more important than safely completing a home inspection. Otherwise, what is the point of being there in the first place? The tips I offer should only be the beginning of your awareness and focus on safety. To be a successful home inspector you must keep your wife from cashing in on your life insurance policy and running off with the pool boy.

CHAPTER THIRTY

LONG-TERM HEALTH AND SAFETY

ABOUT THE TIME I STARTED DOING HOME INSPEC-tions, I remember a conversation a couple older gentlemen were having while playing pool in a bar. One guy was complaining that he couldn't see the balls very well because of all his years as an auto-body repairman—the paint had affected his eyes. The other guy was hobbling around the table, complaining about his knees after decades of installing carpet. Watching these guys—who were a good forty years older than me—has always stuck with me. The world is safe these days. Jobs are safe. In fact, things have never been safer. There are organizations like OSHA, and safety equipment and regulations are everywhere you look.

However, most of these rules and regulations are designed to prevent accidents. There is not much to prevent a worker from wearing out their body. The sad reality is that when some laborer wears out his body, it doesn't cost as much as when there is an accident. That is likely why no one cares as much about what kind of shape a worker is going to be in when they are sixty-five.

Performing home inspections is a pretty safe and healthy job. You spend a lot of time outside, and when you are inside, you are in different houses every day. So, if any one house is toxic, for whatever reason, at least you're not there for long. Still, it's worth paying attention to the long-term risk associated with this job. As I type this I am coming up on my forty-seventh birthday, and though I feel fine now, I could be killing myself for all I know. Look me up in a couple decades, and we'll probably both have a better understanding of the long-term health risks of performing home inspections.

Anyway, without further ado, here are some of the things I do to be as safe as possible.

Always be aware when you are in a house built before 1980. That is roughly the cutoff for some known environmental hazards. Things like lead paint and asbestos should be gone by then so I feel a bit safer running through attics without a mask. I always wear a mask in crawlspaces of any age since I have to get my coveralls and my mask is with my coveralls. Attics? I know I should always wear a dust mask, but have been known to run through attics in newer houses without one if I don't need kneepads and gloves (also in the truck with my mask). Every once in a while I catch a ray of sun as I am standing at an attic access hatch and the amount of dust is pretty alarming. If anything gets me long-term with this job I imagine it will be something affecting my respiratory system. As inspectors we are always running around opening and closing things, and generally stirring up dust. Even aside from the known things like lead paint and asbestos, I have to think all this general dust is not a good thing.

Physically speaking, performing a home inspection is not too demanding, but there is definitely more movement and effort required than pounding on a keyboard in an office cubicle somewhere. Things like shoulders, knees, and backs do get a workout. They also get run down and injured from time to time. I had back surgery about a decade ago—it was technically successful, but I still struggle from time to time. Be ready for lots of bending and twisting over ductwork, lifting a ladder, wrestling tile-covered crawlspace hatches open, and things of that sort. Looking under sinks and just general stooping is a requirement and does run you down at times. My knees are holding up okay, but I have had some problems here and there, and will often wear a soft volleyball-style knee pad under my jeans if I'm nursing a sore knee. I see more of that in my future.

I find that every few years I add another pad of some type to my crawlspace routine. I didn't even wear knee pads when I started, but they are a must now. I've also added elbow pads, which really helps. Gloves with padding on the palms are nice as well. At the rate I'm going, by the time I'm sixty I'll be encased in an air-filled body suit as I crawl around.

I'm always somewhat fighting my weight. If you're one of those people who can finish every meal with a piece of cheesecake and not gain a pound, I am happy for you. I am not so lucky. Most people wouldn't know from looking at me with all my inspection clothes on, but I actually carry a lot of extra weight at times. At six-foot-four I can hide it pretty well. I've managed to shed pounds many times, but it seems the demons always get the best of me and I have an extra fifteen or twenty pounds I'm lugging around. Without a doubt, performing home inspections is easier when you're fit and in good shape. Being able to move easily gives you an advantage over inspectors who struggle. At a minimum, you will be happier doing your job if you are comfortable and fit. In the end, I'd bet a dozen doughnuts the trim healthy inspector is better than the one wheezing and struggling to fit into places. I know this because I have been both.

Be aware of the long-term effects of this job—or any job you have. Our health is all we have, and far too easy to take for granted.

PART FOUR

STAYING EDUCATED AND KEEPING AN EYE ON YOUR COMPETITION

CHAPTER THIRTY-ONE

ONLINE
MESSAGE BOARDS

WITHOUT A DOUBT, THE SINGLE BEST THING I EVER did to become a technically knowledgeable home inspector was to spend time in online message forums. And I mean spend a *lot* of time there. For about five years I would read and answer questions, actively participate in discussions, and post questions for others to answer. Ever since the invention of the internet I've been a message-board junkie. I estimate that over the last nineteen years I've spent more than 5,000 hours on various message boards—many of them about home inspection and construction.

As I sit reading and typing every night my wife probably thinks I am just wasting time staring at my computer. Every once in a while she'll look over as I'm feverishly pounding away at my keyboard and ask, "What are you typing?" The reality is, I've gained more education and knowledge on the great World Wide Web than I could ever have gotten in a classroom. The knowledge and experience that people are willing to give away for free is astonishing.

The topics on these boards span just about anything inspection-related. There are usually separate areas for technical items like electrical and plumbing, and areas for business. You can pretty much choose what you want to learn about. I naturally spend about half my time brushing up on technical topics, and half reading about business operations. There's just nothing better than reading about how other inspectors call things out, what they say about a given problem, what they might not make a big issue about, and how they run their business.

Of course, this knowledge comes at a price. Every message board has a bully or two, and you must often wade through a mountain of BS to arrive at the knowledge you're after. Home inspectors are a funny bunch. We come to this from all walks of life. From hanging out at various conferences and continuing education classes over the years, I've come to the conclusion that a lot of inspectors are the know-it-all type. Many are dying for some authority, and to be able to tell someone what to do. They get out of their truck with a Superman cape on and are going to solve all the world's problems. No matter what shade of blue you say the sky is, Mr. Superman Home Inspector will argue that he knows better. Unfortunately, this type also loves to hang out on message boards. They can always look at a picture and find something that wasn't being asked about.

The good news with message boards is that some really experienced, accomplished, and great folks are there. You just have to have thick enough skin and enough good judgement to get past all the distractions and reach the knowledge worth searching for. It's kind of like picking fruit in an orchard. Many times you walk up to that pear that looks perfect, only to spin it around and find out that it's got a big nasty worm hanging out of the other side. I encourage you to spend as much time as you can on message boards and get to know the people. You will, of course, find plenty of worms. But, the beauty is you find some really great fruit along the way. If you're willing to invest the time and effort, sharing your thoughts and absorbing the information from others will pay off in a big way.

CHAPTER THIRTY-TWO

CONTINUING EDUCATION

WITH LICENSING COMES THE NEED FOR CONtinuing education. A good case can be made that performing home inspections each day is the best continuing education one can get, but I understand the need for something more official. To maintain a license as a home inspector in most states, you need to collect between thirty and forty hours every year or two. The ironic thing is that all you really "must" do is pay for a class and sit through a seminar, and you get a certificate. Over the years, I've seen attendees writing, watching videos, and even sleeping during these seminars. I dream of sneaking a few pictures at key times, using them in my marketing to show just how "dedicated" my competition is to bettering themselves.

The alternative to sitting (or sleeping) through in-person continuing education classes is to gather the necessary hours online. The online thing is usually cheaper, faster, and easier, so why not go for it? Well, getting your lunch at a drive-through is also cheaper, faster, and easier than hunting for real food—but I hope I don't need to explain all the pitfalls with that.

Personally attending continuing education classes is one of the best things you can do to keep an eye on your competition, get invaluable interaction with your peers, and get relevant education to really help you move your business forward. Some of the most beneficial things I have done for my business over the years have come out of attending both local and national inspector conferences. I've found employees, business management programs, field tools and technology, and great friends—all while attending conferences. Keep in mind that these conferences are business expenses, and often, travel and related expenses are tax-deductible (check with your accountant). Being on the West Coast, I particularly like the conferences that occur annually in Las Vegas. Of course, the money lost at the blackjack table isn't usually a tax deduction—but you'll make up for it with all the positive things you learn and people you meet while attending the conference.

In the end, if you are in a state that licenses home inspectors, you are required to obtain continuing education credits. If you aren't in a state that has licensing, you should seek out and attend classes anyway. To become a successful home inspector there is nothing more important than attending conferences to keep an eye and ear on what is going on around you in this ever-changing industry.

CHAPTER THIRTY-THREE

BUILDING CODES AND MANUFACTURER'S SPECIFICATIONS

YOU MIGHT SEE THE TERM "MANUFACTURER'S specifications" thrown around a lot. There are "manufacturer's specifications" with every product you buy. From a pillow to a smartphone to a chainsaw, whoever makes the product has a set of very specific instructions for just how to use it. Truth be told, this isn't because the manufacturers are nice people and care about your enjoyment. No, it's because they are afraid you will die using their product—and that if you do, your family will sue them. The manufacturer of your pillow is afraid you will suffocate yourself in your sleep, the smartphone manufacturer is afraid you will be staring at their product and run a stop sign and crash into a school bus, and the chainsaw manufacturer … well, you get the point.

Yep—every product that goes into a house has a team of lawyers defending it, trying to stay a step ahead of another set of lawyers, who are trying to attack the product for being the most dangerous thing the public was ever exposed to. What does this have to do with houses and inspections? Every building codebook cautions that "manufacturer's instructions must be followed for all products used in the house"—but every set of manufacturer's instructions has a statement that says, "The installation of this product must conform to all local building codes." Basically, nobody wants to take any responsibility for anything (I know, it's a big surprise, given the teams of lawyers). As an inspector, you can leave this legal "hot potato" to someone else, but it's good to know the game that is being played.

You will often run into products installed in a way that seems to violate a building code you think you know. All it takes is a set of manufacturer's specifications saying the installation is okay, and the city inspector is happy to exit stage left, leaving the manufacturer to take any bullets (remember the teams of lawyers?). Yep, everyone just wants to have some plausible deniability. This is why the people who write the codebook put the responsibility on the manufacturers, and the manufacturers put the responsibility on the building code folks. If not for the old saying, "You can't fight city hall," I'm sure our courts would be full of suits between building departments and manufacturers. Luckily, the building departments are largely insulated from liability, so there isn't much for the lawyers to go after. But the manufacturers aren't so lucky. Hence the ever-increasing list of instructions, cautions, and warnings on every product we buy.

The legal world is such a disgusting place these days that I have a hard time not getting aggravated, but knowing the law is hugely relevant to a home inspector. In particular, understand how the building code and manufacturers of products relate to one another. This will come up for you at some point during your inspections—especially if you try to take on the role of Superman inspector and solve all the world's problems. I can guarantee you will be standing in front of something newly installed and permitted by the local building department that you think is wrong. It well may be wrong according to the building code but allowed per the manufacturer's instructions. Most city inspectors choose their battles, and as long as there is a manufacturer stating a given installation is acceptable, they will cite that and move on. It's up to you how much you want to fight with people about these things. If you fight, at best you will be considered difficult and hard-headed. At worst, you will be considered wrong. In either case, you will piss off clients, and lose future work.

Remember: our purpose as home inspectors is to determine if a given system or item is performing as intended. I don't remember reading anything about tearing apart manufacturer's instructions or dissecting the building code in any standards of practice that I've read. Maybe I need to take a closer look. In the meantime, I highly recommend steering clear of these pissing matches. Nothing good comes from fighting with anyone. Do a good inspection, advise your client per the standards of practice and your contract, and move on—unless, of course, you like fielding constant phone calls from angry agents, confused buyers, and hungry lawyers.

CHAPTER THIRTY-FOUR

ENGINEERING AND THE BUILDING CODE BOOK

THE GREATEST HOME INSPECTOR EVER WOULD memorize the code book—then he would know everything about every house, right?

Wrong! Aside from changes in the building code over the years, that code is really just a set of instructions for building a house. Building to code is economical for builders, since all the number crunching and hard planning work is already done. The problem with just inspecting to the codebook is that *any* part of a house can be designed by an engineer, which means it supersedes the code. Many a home inspector has put their foot deep in their mouth by calling something out as "wrong"—when it was engineered, and is, in fact, just fine.

That's right—things that look completely wrong by every measure you've ever learned can be just fine, as long as an engineer is willing to put their stamp on it.

The real problem we have as home inspectors is that we can't possibly know when something was engineered.

This becomes particularly complicated on older houses, where plans get lost, reliable building department records are harder to come by, and construction methods and technologies change. When you just look at something, it's not always possible to know if it was designed and installed perfectly. This is a hard pill for many inspectors to swallow. A lot of us want a problem to be "black and white." We want a book to point to, so we can say definitively that something is right or wrong. Unfortunately, that's just not possible. The "book" is actually a *library* of books that structural engineers study in school.

If you are a black-and-white type, and must have some rule to point to, being a home inspector could be a bit of a struggle. The reality is that being a home inspector involves a whole lot of gray. You must constantly take in data and process it to make a call. I promise you'll never make the perfect call every time. It's just not possible. You will call things as "wrong" that are actually okay. There are obscure exceptions buried in the codebook that you just won't remember or know. Then, as I just discussed, there are things that are engineered that you think are wrong. And of course there's the other side of the coin: something is put together wrong, and you think it's okay. Calling wrong things right, and right things wrong: both are going to happen. Prepare for it, and be ready to deal with it.

There will be more on this later, but the main point is to always keep in mind how the system works. Just because something is not the way the codebook says it should be does not make it wrong. This concept is tough for a lot of inspectors, who think the key to success is memorizing more and more of the codebook. The real key to success is to embrace the gray area between the black and the white. You also must understand how the engineering of houses and the building code relate to one another.

CHAPTER THIRTY-FIVE

FREE TRAINING AND EDUCATION

EVERYBODY LIKES SOMETHING FOR FREE, right? You'll see people go to ridiculous lengths for a "freebie." Often, the people chasing down the free goods or services aren't hurting for money at all. There's just a magical feeling about getting something without paying for it. You'll see people make total fools of themselves, trying to win a radio contest or wait some ridiculous amount of time in line for a free pile of soggy pancakes on "National Pancake Day."

In any case, here are some things you can regularly get for free as a home inspector: training from licensed tradesmen, a look at their work, and a look at the materials they use in their work.

Most tradesmen are more than happy to talk about what they do. They encounter so few people who are interested in anything they have to say. Their days are filled with people just wanting something that is broken to work again, and then listening to those people piss and moan about the cost. Their customers don't care that the reason their house is hot is because the reversing valve relay on their heat pump failed. They just want the house to be cool again. I can guarantee every time you spend more than five minutes talking to a good tradesman onsite you will pick up some useful knowledge. From correct terminology, to installation concepts you never considered, to the inner workings of systems you deal with every day, the things you can learn are amazing.

So where are these techs when you are doing a home inspection? Sometimes at the house you are inspecting, sometimes working on a house nearby, and sometimes in a new construction development. I have probably gotten ahold of the most tradesmen in big developments of new tract houses. Within a hundred feet, you can usually find every trade used to build the house. I've also gone and hunted one of these guys down after I found something wrong with their work that they'd want to know. I'm not talking about a missing cover plate on an outlet, but something they obviously missed when putting the house together—like a missing float switch or tattletale drain in an attic furnace pan. Or an exterior outlet box with a clipped romex cable coiled up inside. Things get missed during construction, and if you see the electrician working next door and you can save them a trip to the house and a bit of embarrassment, why not?

A tract of homes is essentially an assembly line, but rather than the product rolling by the people doing the assembly, the people move down the line through the product. During construction, a tract of homes is a perfect training ground. You can find houses at every stage of production—site cleared and graded, foundation placed, framing, rough-in plumbing/electrical/HVAC, insulation, roof/siding, finishing up the interior, and finally the finished product. Realistically, you often lose access to the house somewhere along the way when the windows and doors are installed—but they are often unlocked during the day. I've found if you move through a big development with purpose (and don't look like a meth-head) no one will ever question what you are doing there. If this home inspection thing doesn't work out, my backup plan it to become a thief and start an "installed once—never used" appliance business. I know where to get an endless supply of product.

Anyway, go walk around some new housing developments and look at the way these places are put together. This is a common practice among home inspectors. I was once in a development looking around, and a superintendent snuck up behind me and asked, "Are you a home inspector?" I must have looked confused as I confirmed I was, because he next said, "Wow, you guys are always trying to learn. I see one of you out here at least once every couple of weeks."

Another great place for some training and learning is at the local Home Depot. Everyone turns their nose up at this orange "big box" store, but they do have a lot to offer. I understand if you would rather go to a specialty supplier to get your kitchen cabinets, tile, and light fixtures, but Home Depot still has a lot to offer. Big box stores are really just a supply house for the housing tracts. Every part of a house can be found on the shelves. Wander around and go look at the parts before they go onto the assembly line. Aside from the "house parts," you will also find the same contractors you see building the houses. You can try to talk to them at the store, but for some reason they always seem to be in a hurry. I've always figured the boss only pays them for time onsite. Once they are on the clock back at the jobsite they are more than happy to take a few minutes and chat with you. Even if you are not working by the hour, take a few minutes and listen to what these guys have to say. The wisdom they so freely share goes a long way in making you a successful home inspector.

CHAPTER THIRTY-SIX

YOUR COMPETITION

AS THE SAYING GOES, "KEEP YOUR FRIENDS close, and your enemies closer." This definitely applies in home inspection. Just to clarify, I really don't consider my competition to be an *enemy*—but you get the point, I'm sure. I've met hundreds of great guys (and gals) who are fellow inspectors. I genuinely value the friendships I've developed with many of them, and would (and have) helped them out in a pinch. I honestly have a lot of respect for the profession and the difficulties inspectors face. I often fall into an "us against them" mentality, and look to help the inspector first and foremost—even if they are my direct competition.

But as much as I respect the profession and will go out of my way to help my competition, they are still my competition, and my family needs to eat. To be a successful home inspector you must take every chance you have to learn about and see what your competition is doing. I love nothing more than when one of my regular agents calls me with some problem they are having post-inspection. My first words are, "Who was the inspector, and can you send me the report?" I have scooped up many, many valuable tips, tricks, and other pieces of information related to my competition just from listening to agents talk about their work.

Another great way to "spy" on your competition is to join a local inspector association. Even if you are just a fly on the wall you can get your money's worth. Being an inspector is somewhat of a lonely plight—you are alone in the field day after day. I often compare and contrast it to putting together "widgets" on an assembly line. As an inspector, you have no chance to see how others are boxing their widgets. You are a factory, with an assembly line of one person. A local association is one of the few chances you have to sit next to someone else doing the same job, and to really learn if you are doing it right. Of course, if you are not putting your widgets together correctly, there are other ways to tell—you are likely getting sued, or your phone isn't ringing, or both.

Here's one more way to get a good look at your competition: check out their websites. At first this seems too obvious, but I've been shocked at how many of my fellow inspectors have a problem doing it. Maybe I just have no shame. I regularly google my competition and dissect their websites. I once even caught a guy plagiarizing content straight off my site, without bothering to remove my company's name. You'd think someone would have the decency to at least steal from a company out of their market! Anyway, in the online world, your competition has to show a lot of their cards in their website and advertising. Take advantage of that, and keep an eye on what they are doing. You will learn some things.

An observation about my competition: most of them claim that they have performed fifteen inspections per week for the last twenty years, that they are always booked out two weeks in advance, and that they have never had a complaint. Realistically, they are a bunch of liars. Maybe that's a bit harsh, but I don't know how else to label it. I suppose boasting about being busy and great at inspecting is rooted in insecurity and fear, but I've never really understood it. Inspectors have so much to learn from one another by honestly discussing the profession that it's a shame we don't embrace that more, and take advantage of it. I spent two years as the president of our local association, and part of my job as the leader was to fill time when speakers finished early. I'd try to start discussions about how people handle complaints, and would get a room of blank stares. Out of fifty inspectors, only a handful would even admit they had ever had a complaint. But researching dispute resolution history through our local licensing board tells quite a different story—complaints happen to everyone. My main point is to not let your competition's insecurity and bravado discourage you. They regularly go through the same crap you do. Statistics say so.

To be a successful home inspector, you must keep an eye on your competition and the things they are doing. I guarantee you will learn some valuable things and pick up some great tricks of the trade along the way.

CHAPTER THIRTY-SEVEN

LEBRON JAMES AND HOME INSPECTIONS

BOY, DO I HAVE SOME WORK TO DO TO TIE ONE OF the greatest basketball players of all time to home inspections, right? Bear with me, I've got something good here.

I was born in 1972 and as I grew up I watched a lot of basketball with my dad. I got to see Magic Johnson, Kareem Abdul-Jabbar, and Larry Bird battle it out for championships throughout the 80s. As I got a bit older, I got to see superstars like Charles Barkley, Shaquille O'Neal, Kobe Bryant, and the almighty Michael Jordan go head-to-head throughout the 90s. More recently, I've been lucky enough to watch LeBron James dominate the NBA.

When not watching basketball and looking for something more than 80s headbanger rock to listen to, I'll often switch over to talk radio. Usually, sports. For those of you who aren't familiar with talk radio, it's essentially just arguing in between commercials. Whether it's government or politics or traffic or sports, talk radio is just people screaming their opinions at each other. On sports radio, they are often arguing about the greatest player of all time.

So, after watching LeBron James play, and listening extensively to people argue about whether or not he is the best of all time, my opinion is that he isn't close to being the greatest player of all time at any one thing. Yep, Kobe was more of a clutch shooter, Barkley was a better rebounding power forward, Shaq was more physically dominant, Larry Bird was a better pure shooter, Magic was a better passer, and Jordan had much more competitive spirit. So, while LeBron James wasn't the best at any one thing, he was second or third best at everything.

I see a lot of similarities between LeBron James being second or third best to being a successful home inspector. No matter what you do, there will always be someone better at any individual aspect of being a home inspector. You will go to a conference and see an inspector in front of the group who is so polished and well-spoken that he should be making movies in Hollywood. You will hop online to check on your competition and find a guy's website ranks at the top of Google and looks like something put together by a six-figure advertising agency. You will get a copy of an inspection report that looks like a collaboration between a professional photographer, a graphic designer, an English professor, and a team of engineers (the folks actually doing the inspection). You will turn on HGTV and see one of your local inspectors walking some (supposed) buyers through their new house. But just because someone can do one thing well doesn't mean they can do everything well.

Being second best at a lot of things will always win the

race, as opposed to being great at one thing and sucking at everything else. Obviously, scoring baskets wins basketball games, so being accurate and good at throwing the ball through the hoop is important, right? The NBA all-time leader for accuracy is a guy named DeAndre Jordan. DeAndre is a nice player and does some good things, but he's far from a household name, and has never been considered anything close to a superstar. In fact, unless you're a pretty dedicated NBA fan you likely have never even heard of DeAndre.

My point is to urge you not to panic when you run across the Michael Jordan of home inspections. The chances are that you are better than him at some things. I realize my advice flies in the face of all the "be the best, you are number one, stand on top of the mountain" programming that has been shoved down our throats since we were kids. I just think it's important to step back and look at the whole picture as opposed to any one thing when evaluating how you are doing.

One last ironic note on basketball legends and success. After retiring from basketball, Michael Jordan (who is widely considered to be the best player of all time) bought an NBA team, the Charlotte Bobcats (now known as the Charlotte Hornets). Through roughly the first ten years of his ownership, they have been without a doubt the most poorly run team in the NBA. Jordan has made numerous blunders—some catastrophic—with drafting and trading players, hiring coaches, and managing the team. So the next time you see Mr. Hollywood home inspector, pay attention and watch, but don't worry too much. You're likely doing just fine.

CHAPTER THIRTY-EIGHT

EMBRACE TECHNOLOGY

AS THE YEARS TICK BY, I CAN'T HELP BUT NOTICE that the bald spot on my head is getting bigger, and the crawlspaces are getting harder to get through. I also can't help but notice every time I look up that there is some new technology that everyone is using. The saying "You are losing ground by standing still" was first uttered long before our current age of lightning-speed high-tech advancement, but it sure does hold true today more than ever. I'm pretty good at attending conferences, reading online, talking to my competition, and generally watching my industry. That being said, I regularly stumble upon a report, or visit a website from a competitor, or read a post from a savvy inspector on a message board forum, and feel as though I have been asleep for twenty years and just woken up. These moments are scary and jarring. I start to think everything I have worked at for years is in jeopardy of being squashed by some slick new inspector with a great idea.

In any industry, it's important to keep an eye on technology. But this seems especially true in the quickly changing home inspection industry. When I started doing inspections back in the dark ages—around 2000—faxing our computer-generated reports was considered state-of-the-art. Shortly after that, people began to use digital pictures in reports—a total mind-blowing game changer for the industry. Then emailing reports became common. Nowadays, there are infrared cameras, drones, and smartphone-generated reports—complete with embedded video clips. Yes—twenty years ago, inspectors churned out handwritten reports on carbon copy forms, and now we're embedding videos (for you old dogs, that means including a link to a video within a report—you click on it to play a little movie of something that took place at the inspection).

As I mentioned earlier, I often feel as though I get passed by with technology, so I'm likely not the best authority on the subject. But I can promise that to be a successful home inspector you must keep up on what is happening and embrace it. An established inspector can survive for a period after digging his heels in and refusing to evolve—but he or she is solely coasting on their personality and past client base. Their business is kind of dying a very slow death.

Here are a couple relevant stories on the dangers of missing out on technology.

About once a year, my office gets a request from a sweet little old lady pulling off a real estate transaction via fax machine. I try to hide my laughter when she asks me to fax the report over. And if *I* find this funny, I can only imagine what the seller of the house and their agent think.

Then there are the buyers who don't have an email address. Okay, I get it. The year was 2005 and you were one of the last holdouts. You refused to do business on that computer thing. Then it was 2010, and if you wanted to talk to your family or see pictures of your grandkids, you needed to get onboard. Soon, it was 2015, and if you didn't have an email address you pretty much couldn't participate in society. Now, we are quickly approaching 2020, and I just don't see how it's possible to live in today's world without an email address.

For me, the final realization that everyone in the world had an email address was when, out of the blue sometime in 2013, I got an email from my seventy-five-year-old mother. If you lined up all of the people in the country based on their computer knowledge, my mom would be at the end, hanging out with the people who didn't know which part of the computer you were supposed to look at. She honestly once thought my sister was talking about the computer when she was actually discussing mini-blinds for her house. Mom said, "Well, you're always talking about your computer having windows. I just figured it would have blinds too." But I digress.

In 2003, I purchased the company I had been working for. The previous owner had been a bit tech-phobic. He refused to buy into the idea of digital pictures. Despite all of our competition moving in that direction, he refused to bow to the "slick and glossy" side of the industry. This drove me nuts as I saw clients drooling over the technology of pictures in reports. I honestly can't even imagine performing inspections without the aid of pictures now. Trying to explain to someone what termite-damaged wood deep in the corner of a crawlspace looked like was pretty difficult and annoying.

I'm so bad at cleaning and organizing things. My attic is often a mess, as is my garage and my office. I have a habit of just throwing things into bins or onto shelves when they are taken out of service. I should just realize I will never use this equipment again, but for some reason I instinctively think a dot matrix printer will be needed again someday. The upside to holding onto all this useless equipment is that I am constantly reminded of how much easier things have become. Holding onto all this crap also makes great material for my kids to bombard me with questions. There's nothing like a kid born in 2010 trying to make sense of a state-of-the-art 1998 PDA ("personal data assistant," for those too young to know, or old enough to have forgotten). The whole concept of the Yellow Pages gets a pretty good laugh from my kids also. Those were the easy days for sure. All you needed to do was add an "A" to the beginning of the first guy in the book and you were all set. String a cord across your living room and plug the phone in, and AAAAA Aardvark Inspections Incorporated was off and running. It's easy to laugh at, but sometimes I wonder if those times weren't easier in ways?

In the end, whether the old days were easier or harder, better or worse is really irrelevant. The fast-moving technology in the home inspection industry is here to stay. You had better jump on the train or you will be left far behind.

CHAPTER THIRTY-NINE

TERMINOLOGY

LET'S SAY YOU WRITE UP SOME PROBLEMS ON A report. You write:

> "The hot water heater needs earthquake strapping, the drain under the guest bathroom toilet is leaking, the joyces for the entry porch have rotted, there is no CO2 detector, and the three-prong outlet to the right of kitchen sink is ungrounded."

These all seem simple enough, right? It's not like there are tons of errors in this write-up. A handyman would likely take this list and fix things without thinking twice. Still, there *are* errors. To be a successful home inspector, it is important to keep in mind the correct terms as you write reports. Most people don't notice the difference, but you will gain points with those who do.

For instance, a water heater is not a hot water heater. (If the water was hot it wouldn't need to be heated.)

The pipe under a toilet is a waste pipe, not a drain

pipe. Drain pipes are small diameter pipes that enter the main waste line. In general, any piping after a toilet is considered waste. Piping from sinks is considered drain piping until it hits the main. This also defines gray versus black water. Gray is drain water. Black is waste water.

The plural form of a joist is joists, not joyces. I've told more than one contractor that "joyces" are little old ladies with blue hair that smoke long brown cigarettes and play bingo together on Friday night. Joists are what hold up your house.

The last time I checked there is no requirement for a carbon dioxide (CO_2) detector in a house unless you want an alarm to go off when you are drinking too many carbonated beverages (a scary thought). A carbon *monoxide* (CO) detector is what is needed.

Outlets have slots or openings, not prongs. Plugs have prongs.

Yes, a lot of this terminology stuff is splitting hairs, but using the right terms shows people you know what you are talking about. Often your report gets handed off to a plumber or an electrician—intermixing key terms makes them conclude that you don't know what you are talking about. On the other hand, I've gotten calls from electricians and plumbers to ask for help in bidding a job (trying to save time without crawling under a house) and have been complimented on my use of terms. It really does help them understand what is going on, and lets them know they can trust what you write up.

Besides—plumbers and electricians buy houses too. There's nothing worse than standing in front of an electric panel talking about grounding the gas piping, only to realize the buyer is an electrician when he reminds you it's "bonding," not "grounding."

PART FIVE

OTHER ADVICE AND WORDS OF WISDOM

CHAPTER FORTY

OH, THE COMPLAINTS AND PUT-DOWNS FROM AGENTS AND CLIENTS

LET ME START BY SAYING I ABSOLUTELY LOVE most real estate agents and past clients. I have a group of professional, successful agents that regularly refer me business, and I am more thankful for that than I could ever express. Most of my past clients are also great folks, and I have stayed in touch with many. (In fact, I'm married to one, but that's a whole other story.) What follows pertains mostly to agents and clients outside of what I would consider to be the norm.

There are no words that cut to my core like when someone begins a sentence by saying, "You missed ..." I don't think agents and past clients realize just what a slap with a wet towel this is. After all, our whole job is to *not* miss things. To be bluntly accused of missing something is just an attack on my professional skills, as far as I'm concerned. The most troubling part of the accusation is that the attacker has usually come to this conclusion through completely unreliable parties or methods. Also, they've blown past any thought of what conditions might have been present when I was there and made a judgement.

Let's look at how a client comes to accuse us of missing something. Most of the time an item is brought to their attention by a contractor looking to sell them something. Sometimes a buyer or some well-meaning friend or family member sees something they think is wrong. Sometimes another inspector comes up with something at the time of resale. In all of these cases the original inspector is completely defenseless. He or she just gets the reactionary blast of the angry client whose first instinct is that they were misled. In my career, I've found that at least 75 percent of the time the information my past client has received is just flat-out incorrect or inaccurate. Maybe 20 percent of the time there is some truth to the accusation, but it's a gray area at best. The remaining 5 percent of the time I made a genuine mistake. Inspectors do miss things and will make it right—either with their check book, their insurance, or their own sweat and blood.

Here are a couple good stories along these lines.

Once, I arrived at an inspection for a buyer that I had done a job for a few weeks prior. The previous house was on a very steep hill and had significant problems— major water intrusion along the front of the house into the crawlspace, cracked foundation, undermined footings throughout the crawlspace, uneven or unsupported floors. Pretty much every nightmare you can think of. Basically, the house was slowly sliding down a very steep hill. As I typically do with things like this, I took a bunch of pictures, made some descriptive comments, and recommended "a full evaluation of all aspects of the crawlspace and foundation with respect to structurally significant movement and water intrusion by a qualified engineer."

As I met the buyer in the driveway at the second house and made the usual chitchat, I asked the obvious question: "So, did you pass on that last house?"

"Yeah," he replied. "We had an engineer out and he found all kinds of things you missed, so we decided to terminate the transaction."

Huh? All kinds of things I missed? Did the engineer just happen to be driving by? Wow, was I steamed. Being a nice guy, and realizing I was being paid, I just shut my mouth and went about my business. I'm thankful I still have the ability to do that. I know many people don't—and as I get older and crankier I may well lose it, too. (One of my favorite comedians, Ron White, had a great line about this. When he's confronted by the police after drinking, he says, "I may have had the right to remain silent, but I didn't have the ability.")

Another time, I pulled up to an inspection for an agent with whom I had worked in the past. She was with another agent who I'd never met before. The second agent met me at my truck and started lighting into me about a past inspection. "So, have you heard from the lawyers about the house up on Johnson Street?" she asked. "The lawsuit is probably going to be into six figures. They have torn the house apart and found all kinds of problems that you missed."

By this point my heart was racing and I was absolutely horrified. As I gathered my things and made my way into the house, I racked my brain trying to remember the past jobs I had done with this agent and her team.

Meanwhile, she came in behind me and announced to the agent I knew *and* our buyer that "I was just telling Matt about the huge lawsuit up on Johnson Street and all the things he missed."

That's when the agent I knew interrupted her, saying, "Oh, no—that wasn't Matt. That was the guy from ACME Inspections that we used to use."

Needless to say I was relieved. But I was also pretty pissed off. Why would this agent attack me going into an inspection—and in front of buyers? Why would she not make sure of her facts before attacking me? There really are no good answers—just as there are no good explanations for why we get treated so poorly in general at times. I chalk it up to buyers and agents having so much on the line, and being so emotional at inspection time. It's certainly no excuse, but I suppose it's an explanation.

Unlike most things I write about, I don't have any great advice on how to completely prevent situations like these. My best advice is to expect the ridiculous and unexpected, and try not to get too rattled when it happens. If you do home inspections for any length of time you will undoubtedly be confronted with unjust and baseless accusations. Many will come at terrible times, while you're trying to focus on other things. It will be difficult to maintain your composure. To be a successful home inspector you must find a way to accept what you can't change, and get through difficult situations the best you can.

A shot of Jack Daniels comes in handy as well ... *after* the inspection, of course.

CHAPTER FORTY-ONE

AN ICE-CREAM CONE

I HOPE YOU'RE LIKE ME AND LIKE A GOOD STORY every now and then. Here's one with a good lesson that's relevant to our industry and customer service in general.

This was probably sixteen years ago, as I was still in my first couple years doing inspections. It was a rainy Saturday afternoon and I pulled up to my inspection about fifteen minutes early. After working around the outside for a bit, I took shelter under a very small porch overhang. Pretty soon my buyer showed up and joined me under the overhang. We stood there for about twenty minutes together, waiting for the real estate agent to show up and get us into the house. The weather turned even worse—it was absolutely pouring. I remember the buyer well—he was a really nice guy, which was a good thing, since we were standing about six inches apart trying to stay dry. Finally, more than thirty minutes late, the real estate agent rolls up in her gaudy BMW, and as I look closely, I realize she's holding an ice-cream cone.

Now, I don't want to go deep into the physical characteristics of ice cream, but it is a pretty safe bet that thirty minutes earlier, when our inspection was supposed to start, the ice cream was still in a freezer somewhere, with her waiting in line to buy it. It was at that moment that I really learned what caring for your client means. For the 10K or so commission check this lady was getting, she couldn't be bothered to be on time.

Once she graced us with her presence, we of course got the obligatory, "Sorry, I got held up" excuse, with nothing else being said. The story of that agent has a funny ending, though. About every two inspections she would call and complain about me missing some minor problem, or tell me her buyers were really upset about something (curiously, I never heard from any of them). Finally, one day she called and lit into me about something, and I pulled a trick out of my bag that I rarely use.

"You know, Maggie," I said, "it just doesn't seem like you and I working together is a very good fit." The dead silence on the other end of the phone was priceless.

In the end she taught me two important lessons. She taught me what it is to really care for a client. She also taught me to not be afraid to fire someone if it isn't working out.

I suppose she also taught me to opt for a milkshake in lieu of a cone if I'm running late—but that's really never a problem for me.

CHAPTER FORTY-TWO

CHUCK IN
A TRUCK AND
EXPENSIVE ELI

SO YOU SHOW UP TO A HOUSE, RUN AROUND snapping pictures for a few hours, and put together a report with a bunch of problems. What comes next?

As inspectors, we largely don't care, as our work is done. But it's good to at least give the subject some thought. A common question as you are rattling off a list of deficiencies in a house is, "Who do I get to fix that?" Of course, this question is well outside of our scope of work, but in order to appear caring, it's good to be able to offer the folks something.

Finding repair contractors for bids and performing work is likely one of the most difficult thing a real estate agent must do during a transaction. As a group, contractors can be difficult to get ahold of, and difficult to communicate with. They often lack follow-through, and are routinely late, or just fail to show up at all. Ironically, I've found the ones that do the best work are the most unreliable and difficult to deal with. In the middle of a real estate transaction, when there are tens of thousands of dollars on the line for everyone, I'm sure this behavior is particularly frustrating.

Let's dig in a little deeper to the different kinds of contractors. Starting at the low end is the one I like to call, "Chuck in a Truck." Chuck has often had a hard life—it's all he can do to get four bald tires on a truck and keep them rolling in one direction long enough to show up at a house. Chuck usually has little regard for things like building codes, keeping an active business license, and doing a job right. For some reason, Chuck always seems to have a dog.

How does Chuck stay in business? Because he's cheap and will generally get the job done. Chuck works well in a crisis. Often, he's experienced a lot of them over the course of his life and feels right at home when something goes wrong. Got a rabid animal in a crawlspace that you need trapped? No problem, Chuck is your guy. Burst pipe? Gutter hanging from a house? Need a French drain dug across the backyard? Chuck can likely get it done in a couple hours, and for cheap. How about replacing a failed circuit board in a furnace, or properly flashing a window that was installed wrong? Eh—Chuck may not be your guy. But be careful—he doesn't know it and will be happy to try.

At the opposite end of the spectrum from Chuck is "Expensive Eli." Eli drives a shiny new truck (usually no dog), has every tool ever made in the back, and can recite chapter and verse from the building codebook. He doesn't do anything unless it's perfect. I can only imagine what Eli's house looks like. He probably polishes his hammers every night, and has them all lined up and named. Don't get me wrong, Eli is great and will get the job done right. But you are going to pay for it. Eli usually requests that people sit down before he hands them his bid.

So, what does all this have to do with you being a home inspector? Consider this: right after asking who will fix something, people ask the inevitable follow-up: "How much is it going to cost?" This is when you must remember Chuck and Eli, and realize you probably can't anticipate the cost of something within a factor of three times. It's okay to go a bit down the "cost road" with people, but be *very* careful to caution them how much the costs will vary. This is why it is critical for them to gather the information *before* they buy the house.

Here's a personal story to drive home the Chuck versus Eli comparison. My wife informed me she wanted our house to be a different color. I hate painting, so down the rabbit hole of gathering bids we went. I'm pretty sure the first guy was drunk—at 11 a.m., by the way, which gets him extra points. His truck had one of those mini-spare tires that looked like it came off a Toyota Corolla, his dog took a big crap on my lawn, and he wanted $3,000. The second guy drove a truck that probably cost more than my house. He was incredibly well-dressed, without a drop of paint anywhere near him. He didn't believe in sprayers and was going to paint my entire house by hand with a free-range horse-hair brush (no joke, he actually said that). Cost? A mere $12,000.

For some reason I was growing more fond of the drunk with the crapping German shepherd. The next day I was laughing and telling this story to an agent and friend of mine, and she reminded me that her brother was a painting contractor. He showed up in a moderately nice truck with four real tires, had some paint splatters on his clothes here and there, and gave me a bid of $5,500.

"When can you start?" I asked. He did a great job.

Realize the agents and buyers you are working with don't always know how the game of working with contractors is played. You really can help them out. Just because a contractor gives a bid doesn't mean it costs that much, and doesn't mean the scope of work is what he says it is. Encourage people to gather multiple bids. Also, educating people about Chuck and Eli is critical. It always amazes me that people can make it so far in life without understanding how something so simple works. And they are often buying a house much nicer than mine. I sometimes think maybe I should have paid a little better attention and stayed in school longer.

CHAPTER FORTY-THREE

BUYERS LISTEN
BUT DON'T HEAR US

KIDS AREN'T RAISED LIKE THEY USED TO BE. LET me pile on with all the other old cranky bastards and complain about how the world is going to hell, kids have no respect, and things are terrible today.

Actually, I don't have nearly that dim of a view—but there is no denying that we raise our kids differently today than we used to. I heard stories of my grandpa getting taken behind the woodshed before finally running away at fourteen, and my dad has alluded to getting smacked around by his parents on several occasions. I know I definitely got whacked a few times when I was a kid (they were well-deserved, I'm sure).

My kids? Much more than a firm grip on the arm and you are labeled a "child abuser" and will be hauled off to jail. I'm not going to debate whether I think all of this is for better or worse, but I'm sure we can all at least agree that it's different. The result of raising kids with less force and fear is that often consequences are learned much later in life—if at all. Falling off your bike? No problem. Nowadays kids are covered head to toe in pads and safety gear. Talk too much crap to a kid two years older than you? No after-school fight in the alley—the worst you can expect is to be forced into a sit-down to discuss your feelings as a group. Don't do your homework and fail a test? No problem. You will get a chance at taking it again.

Then you grow up and don't read your home inspection contract or report. What's the consequence? Well, there are a lot—but it's no surprise people are surprised. After all, today's generation of first-time buyers were raised under a different set of circumstances than I was. If I didn't read something and proceeded anyway, it was on me. I understood that is the way the world works. Nowadays, everyone expects a safety net. Everyone has had one their whole life. Why would anyone expect the great gravy train to end?

Many times as a home inspector you will find yourself needing to communicate a little reality to some nice kids who just weren't raised to understand the way the world works. Buying a house is a big-kid game. Everyone wants to be a homeowner for five years, double what they paid for the house, and then repeat the cycle until they are living in a million-dollar house. The problem comes when these buyers are hit with the reality that they actually have to spend some of their speculative profits on inconvenient things like roofs, sewer lines, and electric panels. By far, the easiest way to deal with unfavorable information is to just ignore it. Many people do that their whole life and it actually works out okay for them. The rest of us don't have trust funds, and must deal with things, eventually.

As a home inspector, be prepared to grab your buyers by the shoulders (figuratively, of course—we don't have time for sit-downs to talk about feelings) and make them hear you. In between them hanging swatches of paint on the wall and measuring to be sure the eighty-five-inch TV will fit, try to grab your buyers' attention and really communicate some of your larger findings. Also, try to express some of the limitations of your inspection. Remember, many buyers are working under the belief that their trusty old safety net is still with them. I'm comfortable going out on a limb saying that 95 percent of home buyers think someone else would be responsible if, when the seller moves their bookcase on the basement wall, a failed foundation is discovered. They don't particularly care who that someone is—just not them. As far as buyers are concerned, they only signed up for the annual-appreciation-of-value, sell-in-five-years program. They can't afford any repairs. And, of course, they gave you $400 to pick up their safety net and throw it over the house.

To be a successful home inspector, realize what your buyers don't realize. Realize what they are going through, and how fast things are moving. They aren't really hearing you.

And, of course, I'll conclude this chapter with a good story. I was at the beach with my family on a nice Saturday afternoon. About 3 p.m. I looked down at my phone and saw the following email: "Hi Matt—you missed something HUGE on my inspection. Call me ASAP, Steve." Well, this sounded great. I just love friendly emails. This guy didn't even have the decency to leave me an address or the date of the inspection. I was able to pull up the report based on his last name. The house was only about ten years old, and was the type you rarely hear back on—clean inspection, nice neighborhood, well cared for, etc.

So I called to see what was going on. It turned out there was a "huge" lump in one of the bedroom floors. But all the buyer was really interested in was learning about my insurance and where he could pick up his check (of course, he'd already had a contractor look at it and was told that it was going to be at least 10K). As I was talking, I remembered the inspection, and remembered that the house was furnished. As I asked him about the problem, he readily admitted there was a "huge bunkbed" right over the lump spot. *Hmm. Okay. Why the hell are you calling me?* That's what I was really thinking, but I suppose I had to play along.

After getting home from the beach on Sunday I went over to the house, and saw that there was indeed a huge lump in the floor. Honestly, it was so bad it was essentially a trip hazard. There was no way a person could walk in the room and not notice this—unless there was a huge bunkbed over it, of course. The severity actually favored me in this case. I also started remembering the buyer—he was a super-nice, clean-cut, "techie" industry guy from the San Francisco Bay area, probably in his late twenties. Apparently after I did the inspection he had renters move in for a year, and was now selling the house. Someone came over to stage it with furniture and discovered the lump.

I told him he had no case against me. I'm not sure what happened after that. The house sold a couple months after I spoke to him, so it's someone else's problem now. The thing that really stuck with me was the buyer's original questions when he found something wrong. I felt like I could read his mind, and the fact that there was no "safety net" didn't even enter it. It was never a question of whether someone else was responsible. He just needed to be directed to that person.

Wow, kids these days.

CHAPTER FORTY-FOUR

SELLERS ARE A BUNCH OF LYING SCUMBAGS

OKAY, I DON'T MEAN ALL SELLERS, AND MAYBE my label is a bit harsh. But this one really strikes a nerve with me.

When you sell a house, it is required (in most areas, I believe) that you fill out a document called a "seller's disclosure statement." The point of this document is for the seller to communicate important information about the property to the potential buyer. A lot of this stuff is just fine print about having the legal authority to sell the property, about whether or not the house is hooked to a sewer, and things like that. But there is also a large section on the condition of the property. There is particular focus on the function of items in the house, and important things like water intrusion and moisture problems.

I'd say about 10 percent of the time (and I'm being generous here) a seller accurately lists defects with the house. As an inspector, you will routinely find things that just flat-out don't work—ranges, outlets, toilets, garage-door openers, furnaces, and many other things—even though on the disclosure document everything is listed as being in perfect working order.

One of the subjects about which sellers are suddenly stricken with amnesia on the disclosure document is water intrusion. This happens so frequently that I don't have to think back very far to come up with a perfect example. At the very last inspection I performed, everything in the basement was up on blocks or shelves, there were water marks on the wood posts and the foundation walls, and there was a running dehumidifier. And this was in the middle of August. I can only imagine what kind of lake that basement was going to become once it started to rain. So, as I often do, I asked the agent and buyer what the seller had to say on their disclosure. The agent looked so jarred by the question I think he forgot the document even existed. After a flurry of email searches he found it, and informed me that the seller wrote "unknown" in the section asking about water in the basement.

Really? Unknown? Wow! At least have the decency to get the dehumidifier out of there on your inspection day, before lying outright to everyone.

Sellers blatantly lying on their disclosures is so common I think most agents don't even realize it's unethical. As far as they're concerned it's just one more piece of paper they need to cram through the transaction. And this document is usually presented to the seller as such. It likely gets emailed or handed to them with a bunch of other things related to selling the house, and barely gets a mention. Then it's thrown before the buyer of the house via their agent, who doesn't care about it either. After all, why should anyone care what the person who has lived in the house for twenty years has to report? There will be an inspector there for three hours. He will surely be able to detect everything that ever has happened, is happening, or will happen in the future. Yep—that $400 you are giving him is the best value the world has ever seen. (All sarcasm aside, it probably is.)

Another example of sudden confusion for sellers on their disclosure form is "remembering" when permits are needed (there is a specific question about it). You will regularly find projects ranging from a few can lights in a kitchen, to a complete addition put on a house—and the sellers conveniently go for the "unknown" box. Roughly half of the time, they will just try to deny that they even did any work on the house. Once they are confronted with facts they can't escape, they change their argument, and try to plead ignorance. Yep, uncle Hal, who has been a "contractor" for thirty years, and did the kitchen addition and remodel, didn't know you needed a permit to vault a ceiling, move around all of the electrical, move the sink to an island, and relocate the back wall of the house twelve feet into the backyard.

Remember the title of this chapter? Do you still think I'm being too harsh?

As the inspector, you have nobody looking out for you. Don't expect the agents to help you out with the seller's disclosure. Aside from pushing the papers through the appropriate channels, they don't have much liability. The seller? They have huge liability—but most are ignorant of it. I've seen some get stung hard with lawsuits, but it doesn't happen often enough for word to spread and scare sellers into filling out accurate disclosures.

My real beef with sellers failing to fill out an accurate disclosure is that it just leaves us inspectors totally exposed to liability that we shouldn't have. Who do you think the buyers of the dehumidifier in the basement house are going to call in the winter when the basement floods? Probably not the seller. They paid nothing for the disclosure statement they got (ironically, exactly what it is worth) but paid me $400. Yep, the instant you get a dollar for anything in our society, you have just provided an implied warranty and guarantee that goes to eternity. A home inspection is a perfect example of this.

Aside from just pointing out the injustice of sellers not playing fair by disclosing defects with the house they are selling, I do have suggestions for how to minimize your exposure. For example, I'm always interested in learning about the sellers of a house. I don't spend a lot of time on this, but will do a bit of snooping and ask some questions. Mainly, I want to know how long the sellers have been in the house, and what they might have done to it. Agents will often have some feeling about the current homeowner and their situation. You can also readily find past sale dates on internet listing sites like Redfin or Zillow. I also love to look around a seller's garage or basement and see just what types of things they may be working on. You usually won't find coils of wiring and buckets of plumbing fittings unless the homeowners are doing some "home cooking," as I like to call it. All these clues can help you understand the mindset of the people on their way out of the house. Paint cans often have dates from when they were mixed. Determining that the bedroom ceiling with the poorly installed skylight was just repainted a week ago is surely a useful piece of information.

The biggest tool you have to combat sellers who omit information or blatantly lie on their disclosure is to just start talking openly about problems you are finding, and whether or not they have been properly reported. For example, when I see leak stains on a ceiling or an oven control panel that doesn't operate, I'll just ask the agent and buyers what is on the disclosure. I know full well that most of the time it won't be listed, but by asking, I'm leading my clients to the questions they should be asking. I try to stop short of throwing around accusations of false-hood and fraud, but I give people plenty of information to start drawing conclusions on their own. I educate the buyers about what a seller "should" be disclosing (after all, it's right on the form—it's not like I'm making things up here). I've probably pissed off a few agents over the years by doing this, but I don't mind. If they don't care enough about their client to protect them from outright fraud, I can only imagine what level of regard they will hold our relationship in when problems arise.

To be a successful home inspector, realize you are reg-ularly on an island by yourself. Nobody is looking out for you—especially not a seller who probably decided they didn't like you before you even arrived at the house.

CHAPTER FORTY-FIVE

BUY LUNCH

BEFORE FINALLY ARRIVING AT DOING HOME inspections I tried a lot of different jobs to get away from waiting tables. I tried a run at managing a fast-food restaurant, worked as a land surveyor's assistant, worked with a private investigator friend delivering court papers, spent some time doing property management, and even bounced at a trendy druggie dance club that closed down at 4 a.m. I was always up for exploring some new career path.

One night, while I was waiting tables, a guest stopped me and wanted to talk for a bit. He launched into this well-polished routine about how he was starting up a new company in the area and was looking for bright young people just like me. He said he was going to be holding some interviews, and asked if I could come meet with him and his business partner to talk about things.

Wow, was I excited. This sounded like the break I had been looking for. So, I excitedly went to meet these two guys at a restaurant. Well, it turns out the "job" was to jump into some multi-level marketing scheme and try to sell long-distance phone service to everyone I knew or met. They showed me a binder loaded with pictures of successful people living in huge houses making six-figure incomes (keep in mind this was around 1990), working twenty hours a week and taking lavish vacations. They talked about the future, and how much money I'd make, and how they had become so successful.

I was so naïve when I was young! These days, I wouldn't let a guy like that get three words out. But setting my naiveté aside for a moment, the thing that really floored me about the "career" these clowns were pushing on me happened when the bill arrived. I had an ice tea and the guys each had a coke. Not only did they ask me for money to cover my drink, they divided up the bill, splitting it between them. Are you kidding me? You guys allegedly make six figures each, and you're hassling a prospective employee for $1.50 for an ice tea at an "interview"? And you can't even buy one another a Coke? You'd think they could just trade off buying each other drinks as they escorted gullible young men off the plank.

I had similar things happen during other jobs as well. I remember sitting at lunch with my bosses while working as a land surveyor. These guys were partners in the firm and did very well. I remember being surprised by them shaking me down for six dollars for my lunch. The firm billed out sixty dollars an hour for me in the field, and I was getting paid ten. So, for the profit I was producing for them each day, I guess springing for my lunch one time was just too much to expect.

Because of these experiences, and because I feel

obligated as the boss when I hang out with my employees, I *always* buy them lunch, or beer, or whatever else. Similarly, if I'm hanging with an agent or a buyer or pretty much anyone I will reach for the bill. I am not rich by any means, but I do okay and feel incredibly thankful for what I have. Picking up a lunch check or a round of beers at a conference with some other inspectors is such a little amount of money in the big picture. If you want to be a successful home inspector, get a good credit card with a nice rewards program, and take care of the people around you every chance you get. You will be remembered in a good way for doing it, and it's always good to have people on your side. You never know when a small good deed will come back to you tenfold.

The alternative is one of my favorite sayings: "You never know when the toes you are stepping on today are going to be attached to the ass you are kissing tomorrow."

CHAPTER FORTY-SIX

BUY EXPENSIVE WINDSHIELD WIPERS

ONE OF THE BIGGEST REGRETS OF MY LIFE IS that I never got to meet my wife's mother. Unfortunately, she passed about a decade before I met my wife. By all accounts, she was a magical woman, and judging by the daughter she raised, I believe it. My wife and her family immigrated to America, and her mother worked day and night to make a better life for her family.

Among all the great things I've heard about my wife's mother, there's one that has always stuck out to me. With the little money she had when she first arrived to the United States, she always believed in buying good shoes and a good mattress. "You must get a good night's sleep, and you work on your feet all day every day," she was famous for saying. She believed if you used it every day, no matter what you paid, it was worth it. This has always stood out to me, and I still remember it.

I never faced anything nearly as dire as what my wife and her family her went through, but definitely had some trying times. For years, I just figured a flashlight was a flashlight, socks were socks, and windshield wipers were all created equal. As I've gotten older and have a few more dollars in my pocket, I want to stress how much easier it is to be a successful home inspector when you have not only nice tools, but *really* nice tools. A flashlight that costs twice as much is twenty times the tool.

There is nothing worse than standing in some grass behind a house, trying to concentrate, and feeling cold wetness as your "water resistant" shoes fail. Trying to see across a crawlspace with a fifteen-dollar flashlight is never a good thing, to say the least. (I've always been surprised that E&O insurance companies don't provide really good flashlights for their clients—I know I do for mine.) I've often joked that I must have gone wrong somewhere in life when a new pair of kneepads is the highlight of my day. But make no mistake—it always feels great strapping on a new pair. There is nothing like that nice soft sensation when plowing my knee into a concrete footing in a crawlspace, and having a nice thick new piece of foam in between my skin and the cement. And finally, there is nothing worse than finishing up two rough inspections, heading for home on a dark rainy November evening, and having your crappy windshield wipers skip, streak, and drag across your window.

As I said in the introduction of this book, being a successful home inspector is not easy. In order to pull it off, you must get the "low hanging fruit." You must minimize the world's small annoyances so you can put your full attention toward your work. Get a nice camera, get comfortable clothes, get a nice headset for calls while on the road (for god's sake don't use precious time at home on these), get a nice computer, and any other things that help you perform at the top of your game. Most importantly, make my mother-in-law proud, and get a great pair of shoes and a good mattress. You can only hope to be lucky enough to keep using both every day.

CHAPTER FORTY-SEVEN

GET OVER IT

I'M SURE YOU'RE WONDERING, "GET OVER WHAT?" I purposely left that vague in the chapter title. If I had listed it, there's a good chance you would have skipped the chapter. But now you're on your fourth sentence and are committed. So here it is: the "what" I'm referring to is public speaking.

Wait! Don't bail on me yet. I promise I can help (without dumb ideas about picturing the audience in their underwear).

To be a successful home inspector, you are going to have to speak in front of people. There is just no way around this fact. Luckily, the groups are usually pretty small—a couple of home buyers and their agent, and maybe a relative or friend or two. But sometimes the groups aren't small. I once had fourteen people crammed into a small living room, listening intently to my wrap-up at the end of the inspection.

During my entire life before doing inspections—and for roughly the first ten years of inspecting, too—I was absolutely uncomfortable and horrified about speaking in front of people, even small groups at inspections. I suppose I hid it well or I never would have made it through the first ten years. Anyway, I finally just got to the point where I thought, *Dammit, I know enough about houses, I am entitled to my opinion, and what I have to say is educational and worth these people's time listening to me.* It really was almost as though I flipped a switch one day and have never looked back. I guess I just realized that my desire for financial success was greater than my fear of public speaking. I now have no problem standing at the end of a driveway and sharing my opinion of the structure with an engineer, or standing in front of the furnace and telling the HVAC tech what I found, or giving some multi-millionaire buyer an honest assessment of a trophy house. Of course, I don't get out over my skis and make things up or try to embellish my experience. I just give them what I think and know to be correct. Speaking within your education and experience is key to this whole public speaking thing.

So, what about larger groups? Once I became comfortable with talking to small groups, I had to test my newfound confidence and talent on larger groups. I was on the board for my local inspector association around this time, and the older president was retiring in the middle of his term and recommended me to fill his position. Well, the next seminar was in a few weeks and I figured no problem. I've conquered public speaking, right? Wrong! I was scared to death and was certain I was going to pass out in front of thirty-five of my competitors. Luckily, the beginning of the seminar went by quickly. Introductions were made—"Good morning ... thanks for getting out of bed to hang out with us on Saturday morning ... here's our first speaker"—and I got through it. Each time in front of the group that day got a bit easier, and I left feeling okay. The next seminars became progressively easier and I got to the point where I actually enjoyed being up in front of the group. I ended up leading one- to two-hour discussions to fill time when we didn't have presenters for our group.

Since speaking in front of my inspector association I've become certified through my state to give education credits to real estate agents and often present to large groups of them at their offices. Similar to my other experiences, I was terrified the first few times I got in front of the group, but have quickly gotten past that and now really enjoy it. Sure, I still get a few butterflies now and then, but that's just normal. Anyone who says they don't get a bit nervous from time to time before presenting is either lying or devoid of feelings.

My point in all of this isn't to toot my own horn but to plant the seed that you are capable. You just have to get educated enough and confident with what you know—then you will realize that what you have to say is important. The larger point is that to grow your market share and move yourself and your inspection company forward, there is just no way to avoid having to do some public speaking. Trust me—if I found a way to "get over it," you can as well.

CHAPTER FORTY-EIGHT

BE IN A
GOOD MOOD

THERE IS A CONTAGIOUS ASPECT TO BEHAVIOR, and I really try to emulate the successful people I see. Over the years, I have had the privilege of working with some incredibly successful real estate agents. After a while I couldn't help but notice they all have one similar trait: they are *always* upbeat and positive. There isn't a problem that can't be solved, or a situation that can't be figured out. 80 percent of the agents I listen to as I'm inspecting do it wrong. They get hung up on petty details and spend too much energy complaining. I hear it daily as I am inspecting. The 20 percent who do it right (who sell 90 percent of the houses, by the way) just do things more magnanimously. It's almost something you can't teach. Their voice and body language just have a different cadence from the others.

Being the observant young fellow I was when I started inspecting houses, it didn't take long to draw the correlation between these upbeat, positive folks and the really,

really nice cars they always seemed to be driving. I quickly learned that being positive all of the time was absolutely critical to growing my brand as a home inspector. No matter what people throw at you, find a way to get through the work in front of you, take care of your client, and move on.

Don't get me wrong—I have plenty of terribly negative things to say about my work. But they are only heard by my poor wife once I'm at home, or by some friends over a beer. As far as my inspecting goes, I am completely "on camera" while in front of clients, while I am doing an inspection, or while I am otherwise in view.

This brings up another point: every time you are inspecting a house you are in an unfamiliar house. With today's technology, I can promise you are being recorded with video or audio (or both) much more often than you realize. When I train my employees, I warn them to work as if they are being watched at all times. The worst thing anyone should ever catch you doing on camera is picking your nose. Don't snoop, don't go through things, obviously don't take anything (whether there is a camera on you or not). Work as though you are being watched, and talk to your clients as though everyone is listening. Again, remain upbeat and positive. I can easily tear a house to shreds factually while remaining positive and upbeat. The sellers surely won't like the news I'm delivering, but it is factual—and I am just the messenger, as the saying goes. Agents really appreciate a positive delivery and an attitude that every problem is manageable.

CONCLUSION

A HOUSE IS AN AMAZING AND COMPLICATED PIECE of construction. I'd venture to say there isn't a person alive who knows all the pieces and parts that are needed to put one together. To me, there is nothing better than a house. It's the place we raise our kids, visit with family, and relax after a hard day's work. It's also where we spend our lives talking with our spouse, and planning so that someday we can get a better house, or a house in a place we consider to be more desirable for whatever reason.

As I read and reread everything I have written, I fear it comes off as too cynical toward my profession at times. I really don't mean it that way. I just want those who are thinking of entering the profession to go in with their eyes wide open. I also want to tell people already in the profession they are not in it alone. Home inspectors all go through the same things, whether we are willing to admit it or not. There will be complaints, angry buyers, angry real estate agents, angry sellers, and contractors accusing us of being total fools. Don't let it discourage you. It's all part of the process. In the end, there are way more good days than bad. Try hard to remember the good days, and forget the bad ones once you've learned from them.

I've never held anything close to a job in a trade associated with building a house. Sounds scary, right? Trust me, it scares me too. I make a good living every day criticizing and critiquing the work of skilled, hardworking tradespeople, but I've never walked ten feet in their shoes—let alone a mile, as the saying goes. Something my old business partner Bob once told me applies here: "Just because I know my watch is broken doesn't mean I know how to fix it." This concept is a very big part of doing home inspections. As home inspectors we don't have all the answers. We raise a lot more questions than we ever answer. This will annoy agents, sellers, and buyers to no end. But it's the way the game is played. We are there to be the eyes and ears of our homebuyers. We are there to use our knowledge and experience to help them make a more informed decision about whether to buy a house or not. We can't have all of the answers. It's just not possible.

I know I view the profession differently than a lot of my competition and others in the industry, but that's okay with me. At the end of the day I am a successful home inspector.

ACKNOWLEDGEMENTS

EVERY KEYSTROKE I MAKE ON THIS LAPTOP—whether to write inspection reports or to put together this book—is for my family. Without them, none of the things I do would have any purpose. My oldest son, Jacob, was born a few months before I embarked on my career as a home inspector, and it was spending time with him that motivated me to find a better way to earn a living on this earth. My wife Krysta and our three other children—Jessica, James, and Jonathan, along with Jacob—drive me each and every day to succeed. They put up with my crankiness when we get inspection complaints, and throughout all the ups and downs of life running a home inspection company. They have also been incredibly patient and understanding as I roll in the driveway every night smelling of crawlspace funk. I always remember the term Jessica coined when she was three. She said, "Daddy, you smell barfy."

I also want to put out a big thank you to Andrew

Drukin, Masha Shubin, and all the great folks at Ink-water Press for their help editing, designing and putting this book together. I am very grateful to have found such a fantastic group of people to take my incoherent ramblings and ideas and turn them into something fit for print.

As for how I have become a successful home inspector, I have been incredibly lucky to be surrounded by great people over the years. My business partner of thirteen years, Bob Buckley, helped me and taught me many things about home inspections, business, and life. I am incredibly grateful for the time we shared together. I also want to thank Bill Hendrix for all his help initially training me nineteen years ago, and for his continued willingness to answer all my questions from his position in the Happy Valley Building Department. I have also been very lucky getting to know other wonderful people over the last nineteen years in my career. My co-workers, employees, clients, and even competition have all been a great help to me.

I owe the largest debt of gratitude to Don and Lilian Crawford, for taking a big chance and hiring an inexperienced, baby-faced twenty-seven-year-old with "a lot of really good questions." There isn't a day that goes by that I don't remember and use the lessons they taught me.

NUTS AND BOLTS

I KNOW I SAID I WASN'T GOING TO GET INTO TOO much technical material in this book, but I've got to give you something. We inspectors really can't help but talk about this stuff. Walking around looking for problems isn't something you can easily turn off. I'm regularly caught staring up at some roof trusses in a gymnasium when my wife elbows me because my kid just shot a basket. And having a home inspector over to your house for dinner can result in more than you ever wanted to know about your house.

To be a successful home inspector, you absolutely must be on top of your game with respect to how houses are currently put together, and how they have been put together for the last century. Just being a good handyman or having people praise you when you glue a loose toilet in place isn't nearly enough. You need to be able to walk the aisles at Home Depot and stop anywhere and know *exactly* what you are looking at and what it is used for.

In no particular order, and making no attempt to cover all material in a section or the house as a whole, here is a list of a few of the things that rattle around in my head as I wander into "the office" each morning.

ROOFS:

- rake flashing
- drip-edge flashing
- head flashing
- valley flashing
- step flashing at the chimney
- counter-flashing at the chimney
- furnace/water heater vent pipe flashing
- step flashing at siding to roof seams
- moss/leaves/debris
- overhanging trees
- roof deck feel underfoot
- shingle type/age/performance
- shingle layout and exposure
- starter course
- hip/ridge shingles
- skylight installation and flashing
- wood shakes
- concrete tiles
- clay tiles
- torch-down roofing
- TPO roofing
- EPDM roofing
- vinyl membrane roofing
- metal shingles
- metal panels

- different flashing methods for penetrations and edges of each roof type
- standing seam metal roof installation versus metal shed roof material
- Underlayment material—15# felt, 30# felt, synthetic
- must verify the pitch of the roof is compatible with the materials used
- must verify how many layers of roofing are present (take note if multiple roof are present, as the added weight could cause broken rafters—remember to check closely when in the attic).

All of this is just for the part of the roof that is outside—once inside, looking for leak stains is one of the most important parts of the inspection.

ELECTRICAL:

- service to the house
- overhead drop height above a walkway
- height above a driveway
- meter base
- access to equipment
- main panel bonding/grounding
- sub-panel bonding/grounding
- outbuilding service, bonding/grounding
- service wire amp rating
- main breaker rating
- panel amp rating
- be aware of the "bad" brands (Federal Pacific, Zinsco)
- check breaker to wire size correlation
- openings in the panel for oxygen or rodents

- excess heat at connection points
- cables and wires trimmed/routed/marked/secured properly in and around the panel
- water/moisture presence inside the panel
- verify working space in front of the panel
- outlet and switch height and clearance
- no outlets or connections near garage floors
- cables bundled over long distances must be de-rated
- smoke and CO detectors in the house
- GFI outlet location and functionality
- outlet wiring
- ungrounded outlets
- reverse polarized outlets
- two-slot outlets with grounded boxes
- two-slot outlets with ungrounded boxes
- improperly wired three-way switches, four-way switches, five-way switches (how do electricians actually wire these anyway?)
- wire junctions hanging in space
- cables not clamped near boxes
- boxes missing grommets to protect cables at entry points
- proper lighting at stairways, exterior doors, hall-ways, and bedrooms
- no bulb fixtures in closets
- worn/broken outlets
- verify all equipment outdoors is the correct type
- damp location versus wet location
- service disconnects for equipment—line of sight between disconnects and equipment
- switches and outlets missing cover plates

- wires hanging from walls and ceilings (are they live? an unfinished project? old wiring taken out of service?)
- wiring types—knob and tube, bx cable, NM cable, NM-B cable, UF cable, metal conduit used as ground
- aluminum wiring for service entry, aluminum wiring for 240V circuits, aluminum wiring for light/outlet circuits (two out of three of those aluminum uses are okay, but one isn't—you'd better know which because if you call it wrong your phone will blow up with an angry agent, client, electrician, or maybe all three)

KITCHENS:

- Test outlets
- check for gfi protection
- hot and cold water
- functional water flow
- piping materials (supply and drain)
- lights
- dishwasher drain hose high loop or air gap
- dishwasher secured to its cabinet
- dishwasher physical condition and operation (must remember to check for leaks after cycle has completed—not doing so cost me $1000 once)
- garbage disposal operation (is the inside rusted? are there cracks or leaks in the housing?)
- disposal electrical supply
- drain piping from kitchen sink: islands can be tricky—usually should have an air-admittance valve, sometimes a high loop vent

- instant-hot water dispenser
- cooktop burners
- range anti-tip bracket for free-standing units
- clearance above range to exhaust hood or microwave
- clearance at side walls
- exhaust fan (outdoor venting required for gas—this is a newer rule, though, so an older installation could be grandfathered)
- exhaust fan ducting material/routing/cleanliness
- oven bake/broil, physical condition and keypad operation
- refrigerator icemaker leakage
- trash compactor
- cabinet and drawer condition and operation
- countertop installation and condition
- floor coverings
- walls/ceilings/windows
- no combustion air for gas appliances can be drawn from kitchens
- no HVAC system return air duct openings in kitchens

These are just some of the things you must know in some areas of a house. This information should either scare the hell out of you or give you confidence that you are on your way to being a successful home inspector. All of the advice and concepts in this book will be of no use if don't *really* know your way through every nut and bolt in a house.

A final piece of advice: Your entire life you have been throwing away the directions every time you open something. Now is the time to start reading them. You will regularly find yourself researching a product installation and manufacturer's instructions to determine if something is installed correctly. Sorry to sentence you to such a dim fate! But look on the bright side—you've had a pretty good run ignoring and throwing away instructions up to this point.

CPSIA information can be obtained
at www.ICGtesting.com
Printed in the USA
LVHW012039100620
657787LV00010B/738

9 781629 016849